Transportation Engineering

Transportation Engineering

Alfredo Adkins

Willford Press
www.willfordpress.com

Published by Willford Press,
118-35 Queens Blvd., Suite 400,
Forest Hills, NY 11375, USA

ISBN: 978-1-68285-736-6

Cataloging-in-Publication Data

Transportation engineering / Alfredo Adkins.
p. cm.
Includes bibliographical references and index.
ISBN 978-1-68285-736-6
1. Transportation engineering. 2. Civil engineering. 3. Transportation. I. Adkins, Alfredo.
TA1145 .T73 2019
629.04--dc23

For information on all Willford Press publications
visit our website at www.willfordpress.com

Willford Press

Printed in China.

Contents

Preface

Transportation engineering is the branch of engineering that is concerned with the design, planning, construction, operation and management of any mode of transportation for the easy, safe, efficient and economically viable conveyance of people and goods. Traffic engineering is an important aspect of transportation engineering. Modern technologies involve advanced traffic control systems, traveler information systems, intelligent transportation systems and vehicle infrastructure integration. User interface of road signals, signs and markings and driver-vehicle interface are important aspects of this field. Transportation engineering is an upcoming field of science that has undergone rapid development over the past few decades. This book covers all significant studies, theories and applications of this field. Coherent flow of topics, student-friendly language and extensive use of examples make this textbook an invaluable source of knowledge.

To facilitate a deeper understanding of the contents of this book a short introduction of every chapter is written below:

Chapter 1- The different modes of conveyance of goods, animals and people are termed as modes of transport. The common modes are air, water and land transport. This is an introductory chapter, which will introduce briefly all the significant modes of transportation, such as transport network, land transport, maritime transport, air transport and intermodal passenger transport, besides others.

Chapter 2- Transportation engineering refers to the application of technology and engineering to the operation, planning and management of the facilities or infrastructure required for any mode of transportation. The aim of this chapter is to explore the fundamental concepts central to this field, such as urban and public transport planning, traffic analysis zone, access management, bicycle transportation planning and engineering, etc.

Chapter 3- Highway engineering is a sub-field of civil engineering that is concerned with the planning, design, construction and management of roads, tunnels and bridges for the efficient transportation of people and goods. This chapter has been carefully written to provide an easy understanding of the varied aspects of highway engineering like geometric design of roads or highways, road surface and road infrastructure.

Chapter 4- Traffic control and management deals with the planning and control of transport services for the effective movement of vehicles and freight. In order to completely understand traffic control and management, it is necessary to understand the techniques and methods related to it. The following chapter elucidates the varied aspects of traffic flow, traffic congestion, traffic reporting, road traffic control, traffic optimization, active traffic management, etc.

I would like to share the credit of this book with my editorial team who worked tirelessly on this book. I owe the completion of this book to the never-ending support of my family, who supported me throughout the project.

Alfredo Adkins

1

Modes of Transportation

The different modes of conveyance of goods, animals and people are termed as modes of transport. The common modes are air, water and land transport. This is an introductory chapter, which will introduce briefly all the significant modes of transportation, such as transport network, land transport, maritime transport, air transport and intermodal passenger transport, besides others.

Transport

Transportation is one of our most basic human needs. Without it, we would die pretty quickly. Transportation is the act of moving something from one place to another. We need transportation because all of life's necessities and pleasures can't possibly fit within the reach of our static bodies from birth to death. We either have to move ourselves to things or have things moved to us. Transportation gets food to our mouths, glycogen to our toes, rabbits to other rabbits to do rabbit things, and people to their boyfriend's gallery openings. This is essentially all transportation is. It doesn't matter how it happens.

Transportation concerns the movement of products from a source—such as a plant, factory, or workshop—to a destination—such as a warehouse, customer, or retail store. Transportation may take place by air, water, rail, road, pipeline, or cable routes, using planes, boats, trains, trucks, and telecommunications equipment as the means of transportation. The goal for any business owner is to minimize transportation costs while also meeting demand for products. Transportation costs generally depend upon the distance between the source and the destination, the means of transportation chosen, and the size and quantity of the product to be shipped. In many cases, there are several sources and many destinations for the same product, which adds a significant level of complexity to the problem of minimizing transportation costs. Indeed, the United States boasts the world's largest and most complex transportation system, with four million miles' worth of roads, a railroad network that could circle the earth almost seven times if laid out in a straight line, and enough oil and gas lines to circle the globe 56 times.

The growth of the ability—and the need—to transport large quantities of goods or numbers of people over long distances at high speeds in comfort and safety has been an index of civilization and in particular of technological progress.

Transportation allows economic development to occur because it's a basic prerequisite for anything human. Neither could our economy exist without air, sunlight, or an expectation of immediate bodily safety.

Transportation planners generally define a person's ability to transport in two ways: mobility and access. Mobility is the extent to which you're able to physically move.

Access is the extent to which you're able to get to the things you want to get to, or have them brought to you. The important distinction is between method and end. Access is the end to which mobility is the means. (In almost every case).

Modes

Modes of transport or types of transportation is referred to a combination of networks, infrastructures, vehicles and operations. These include walking, the road transport system, rail, ship transport and modern aviation. Different modes of transportation have emerged over time, basically there are five (5) modes of transportation.

Different Modes of Transport

1. Road transport: road transport exist in all parts of the world, this involves the use of motor vehicles (cars, lorries, buses, bicycles, and trucks). There are various types of roads according to size and functions, some roads are tarred while others are not. The best of these roads are the modern roads which links major towns. Road transport when compared with other modes of transportation is more flexible. It is relatively cheaper and faster. Road transport has a high capacity of carrying goods over short distances. Maintenance is one of the major disadvantages of this mode of transport.

2. Railway transport: railways were developed during the period of industrial revolution in the 19th century, these was partly for political reasons and for economic reasons. In many countries, they were built especially to penetrate isolated regions and help promote political unity. The major advantage of railway transport includes provision reliable services. It has ability of conveying heavy and bulky goods; it is also very cheap, safe and comfortable for passengers over a long distance.

3. Water transport: water transport is very important because it is the cheapest way of transporting bulky goods over a long distance. In the world, there are two major types of water transport namely: Inland water transport and ocean water transport. Inland water transport- this is the system of transport through all navigable rivers, lakes and man-made canals. Many large rivers in different parts of the world are used by ships and barges for transportation; the main rivers where inland water transport are important are the Rhine and Dambe in Europe, the Zaire in Africa, the Nile in Africa, the Mississippi in USA etc. However, ocean waterways carry a lot of the worldâ€™s trade, majority of the bulky goods, materials and passengers pass through ocean waterways from one country to another at the cheapest cost.

4. Air transport- air transport is the newest means of transport, it was introduced in 1903 but developed into full means of transporting people and goods in 1930s. The greatest of the air transportation started after the Second World War (WW11). This mode of transportation can be used for both domestic and international flights.

5. Pipeline transport: this system of transportation involves the use of hollow pipes in the transportation of water, crude oil, (petroleum) and gas. This mode of transportation is safer than using tankers or trailers in the transportation of these liquids.

Other Modes of Transportation Include

- Animal-powered transport, which is mostly referred to as beast of burden. It is the oldest means of transportation; this usually involves the use of animals for the transportation of people and goods. Humans may ride some of the higher animals directly, or harness them. Example of such animals used for transporting humans and goods include camel, horse, donkey, elephant, and giraffe.
- Human powered transport: this is another form of transport, which people, goods or both transported from one place to another using human muscle-power, in the form of walking, running and swimming. Modern technology has allowed machines to take over human power. Human-powered transport remains popular for reasons of cost-saving, physical exercise, leisure, and environmentalism; it is sometimes the only type available, especially in underdeveloped or inaccessible regions.
- Spaceflight is a means of transport that moves out of Earth's atmosphere into outer space by means of a spacecraft. While large amounts of research have gone into technology, it is not commonly used except to put satellites into orbit, and conduct scientific experiments.
- Cable transport: cable transport is a broad class of transport modes that have cables as the foundation for transporting goods or people, often in vehicles called cable cars. The cable may be driven or passive; items may be moved by pulling, sailing, sliding or by drives within the object being moved on cableways, this is another means of transport use in mountain. The use of pulleys and balancing of loads going up and down are common elements of cable transport.

Elements

Operation

Private transport is only subject to the owner of the vehicle, who operates the vehicle themselves. For public transport and freight transport, operations are done through private enterprise or by governments. The infrastructure and vehicles may be owned and operated by the same company, or they may be operated by different entities. Traditionally, many countries have had a national airline and national railway. Since the 1980s, many of these have been privatized. International shipping remains a highly competitive industry with little regulation, but ports can be public-owned.

Medical Transport

An ambulance is a vehicle used to transport people from or between places of treatment, and in some instances will also provide out-of-hospital medical care to the patient. The word is often associated with road-going "emergency ambulances", which form part of emergency medical services, administering emergency care to those with acute medical problems.

Air medical services is a comprehensive term covering the use of air transportation to move patients to and from healthcare facilities and accident scenes. Personnel provide comprehensive prehospital and emergency and critical care to all types of patients during aeromedical evacuation or rescue operations, aboard helicopters, propeller aircraft, or jet aircraft.

An ambulance from World War I

Economic

Transport is a key component of growth and globalization, such as in Seattle, Washington, United States.

Transport is a key necessity for specialization—allowing production and consumption of products to occur at different locations. Transport has throughout history been a spur to expansion; better transport allows more trade and a greater spread of people. Economic growth has always been dependent on increasing the capacity and rationality of transport. But the infrastructure and operation of transport have a great impact on the land and is the largest drainer of energy, making transport sustainability a major issue.

Due to the way modern cities and communities are planned and operated, a physical distinction between home and work is usually created, forcing people to transport themselves to places of work, study, or leisure, as well as to temporarily relocate for other daily activities. Passenger transport is also the essence of tourism, a major part of recreational transport. Commerce requires the transport of people to conduct business, either to allow face-to-face communication for important decisions or to move specialists from their regular place of work to sites where they are needed.

Planning

Transport planning allows for high utilization and less impact regarding new infrastructure. Using models of transport forecasting, planners are able to predict future transport patterns. On the operative level, logistics allows owners of cargo to plan transport as part of the supply chain. Transport as a field is also studied through transport economics, a component for the creation of regulation policy by authorities. Transport engineering, a sub-discipline of civil engineering, must take into account trip generation, trip distribution, mode choice and route assignment, while the operative level is handled through traffic engineering.

The engineering of this roundabout in Bristol, United Kingdom, attempts to make traffic flow free-moving.

Because of the negative impacts incurred, transport often becomes the subject of controversy related to choice of mode, as well as increased capacity. Automotive transport can be seen as a tragedy of the commons, where the flexibility and comfort for the individual deteriorate the natural and urban environment for all. Density of development depends on mode of transport, with public transport allowing for better spatial utilization. Good land use keeps common activities close to people's homes and places higher-density development closer to transport lines and hubs, to minimize the need for transport. There are economies of agglomeration. Beyond transportation some land uses are more efficient when clustered. Transportation facilities consume land, and in cities, pavement (devoted to streets and parking) can easily exceed 20 percent of the total land use. An efficient transport system can reduce land waste.

Too much infrastructure and too much smoothing for maximum vehicle throughput means that in many cities there is too much traffic and many—if not all—of the negative impacts that come with it. It is only in recent years that traditional practices have started to be questioned in many places, and as a result of new types of analysis which bring in a much broader range of skills than those traditionally relied on—spanning such areas as environmental impact analysis, public health, sociologists as well as economists—the viability of the old mobility solutions is increasingly being questioned.

Environment

Traffic congestion persists in São Paulo, Brazil, despite the no-drive days based on license numbers.

Transport is a major use of energy and burns most of the world's petroleum. This creates air pollution, including nitrous oxides and particulates, and is a significant contributor to global warming through emission of carbon dioxide, for which transport is the fastest-growing emission sector. By subsector, road transport is the largest contributor to global warming. Environmental regulations in developed countries have reduced individual vehicles' emissions; however, this has been offset by increases in the numbers of vehicles and in the use of each vehicle. Some pathways to reduce the carbon emissions of road vehicles considerably have been studied. Energy use and emissions vary largely between modes, causing environmentalists to call for a transition from air and road to rail and human-powered transport, as well as increased transport electrification and energy efficiency.

Other environmental impacts of transport systems include traffic congestion and automobile-oriented urban sprawl, which can consume natural habitat and agricultural lands. By reducing transportation emissions globally, it is predicted that there will be significant positive effects on Earth's air quality, acid rain, smog and climate change.

Mode of Transport

The physical movement of goods across borders, or transport of goods, can be undertaken using one or more modes of transport.

The different modes (air, sea, road, rail, inland waterways) have varying processes and information requirements. This partly stems from different infrastructures, different capabilities for handling larger or smaller amounts of cargo, but also from different international, national and even local regimes for a specific mode.

Every mode of transport has specific characteristics in terms of conditions of transportation, performance and legal frameworks. When using long-haul transportation (ocean ships, intercontinental airflights, long distance rail) the pre- and on-carriage is usually through another mode of transport, often road. This intermodal and multimodal transport can lead to complications and trade facilitation issues such as the use of waybills for other modes of transport.

Transport by air and sea usually includes transport by other modes of transport for pre- and on-carriage (road, rail, inland waterways). Multimodal transport consists of the use of more than one mode of transport, but also involves its own equipment, particularly in rail-road movements through specific equipment that can be transferred from truck to wagon.

As documentary requirements differ from mode to mode, transport involving more than one mode can lead to additional changes in documentation. Sometimes the use of transport documents for one mode is allowed (or excepted) in another mode, for example the "flying trucks" that use airway bills.

Animal-powered

Animal-powered transport is the use of working animals for the movement of people and goods. Humans may ride some of the animals directly, use them as pack animals for carrying goods, or harness them, alone or in teams, to pull sleds or wheeled vehicles.

Air

Air France Airbus A318 landing at London Heathrow Airport

A fixed-wing aircraft, typically airplane, is a heavier-than-air flight vehicle, in which the special geometry of the wing generates lift. Fixed-wing aircraft ranges from small trainers and recreational aircraft to j j large airliners and military cargo aircraft. For short distances or in places without runways, helicopters can be practical. (Other types of aircraft, like autogyros and airships, are not a significant portion of air transport.)

Air transport is the fastest method of transport, Commercial jets reach speeds of up to 955 kilometres per hour (593 mph) and a considerably higher ground speed if there is a jet stream tailwind, while piston-powered general aviation aircraft may reach up to 555 kilometres per hour (345 mph) or more. This celerity comes with higher cost and energy use, and aviation's impacts to the environment and particularly the global climate require consideration when comparing modes of transportation. The Intergovernmental Panel on Climate Change (IPCC) estimates a commercial jet's flight to have some 2-4 times the effect on the climate than if the same CO_2 emissions were made at ground level, because of different atmospheric chemistry and radiative forcing effects at the higher altitude. U.S. airlines alone burned about 16.2 billion gallons of fuel during the twelve months between October 2013 and September 2014. WHO estimates that globally as many as

500,000 people at a time are on planes. The global trend has been for increasing numbers of people to travel by air, and individually to do so with increasing frequency and over longer distances, a dilemma that has the attention of climate scientists and other researchers, the press, and the World Wide Web. The issue of impacts from frequent travel, particularly by air because of the longer distances that are easily covered in one or a few days, is called hypermobility and has been a topic of research and governmental concern for many years.

Land

Land transport covers all land-based transportation systems that provide for the movement of people, goods and services. Land transport plays a vital role in linking communities to each other. Land transport is a key factor in urban planning. It consists of 2 kinds, rail and road.

Rail

German ICE 1 on the Nuremberg–Munichhigh-speed railway

Rail transport is a means of conveyance of passengers and goods by way of wheeled vehicles running on rail track, known as a railway or railroad. The rails are anchored perpendicular to railroad train consists of one or more connected vehicles that run on the rails. Propulsion is commonly provided by a locomotive, that hauls a series of unpowered cars, that can carry passengers or freight. The locomotive can be powered by steam, diesel or by electricity supplied by trackside systems. Alternatively, some or all the cars can be powered, known as a multiple unit. Also, a train can be powered by horses, cables, gravity, pneumatics and gas turbines. Railed vehicles move with much less friction than rubber tires on paved roads, making trains more energy efficient, though not as efficient as ships.

Intercity trains are long-haul services connecting cities; modern high-speed rail is capable of speeds up to 430 km/h (270 mph), but this requires a specially built track. Regional and commuter trains feed cities from suburbs and surrounding areas, while intra-urban transport is performed by high-capacity tramways and rapid transits, often making up the backbone of a city's public transport. Freight trains traditionally used box cars, requiring manual loading and unloading of the cargo. Since the 1960s, container trains have become the dominant solution for general freight, while large quantities of bulk are transported by dedicated trains.

Road

A road is an identifiable route of travel, usually surfaced with gravel, asphalt or concrete, and supporting land passage by foot or by a number of vehicles.

The most common road vehicle in the developed world is the automobile, a wheeled passenger vehicle that carries its own motor. As of 2002, there were 591 million automobiles worldwide. Other users of roads include motorcars, motorcycles, buses, trucks, bicycles and pedestrians, and special provisions are sometimes made for each of these. For example, the use of bus lanes give priority for public transport, and cycle lanes provide special areas of road for bicycles to use.

Motorcars offer high flexibility, but are deemed with high energy and area use, and the main source of noise and air pollution in cities; buses allow for more efficient travel at the cost of reduced flexibility. Road transport by truck is often the initial and final stage of freight transport. Currently there is discussion from Elon Musk about a vacuumed sealed train that can travel up to 720 MPH.

Traffic on the Eastshore Freeway (Interstate 80) near Berkeley, California, United States

Water

Water transport is the process of transport that a watercraft, such as a barge, boat, ship or sailboat, makes over a body of water, such as a sea, ocean, lake, canal or river. If a boat or other vessel can successfully pass through a waterway it is known as a navigable waterway. The need for buoyancy unites watercraft, and makes the hull a dominant aspect of its construction, maintenance and appearance. When a boat is floating on the water the hull of the boat is pushing aside water where the hull now is, this is known as displacement.

In the 1800s, the first steamboats were developed, using a steam engine to drive a paddle wheel or propeller to move the ship. The steam was produced using wood or coal. Now, most ships have an engine using a slightly refined type of petroleum called bunker fuel. Some ships, such as submarines, use nuclear power to produce the steam. Recreational or educational craft still use wind power, while some smaller craft use internal combustion engines to drive one or more propellers, or in the case of jet boats, an inboard water jet. In shallow draft areas, hovercraft are propelled by large pusher-prop fans.

Although slow, modern sea transport is a highly effective method of transporting large quantities of non-perishable goods. Commercial vessels, nearly 35,000 in number, carried 7.4 billion tons of cargo in 2007. Transport by water is significantly less costly than air transport for transcontinental shipping; short sea shipping and ferries remain viable in coastal areas.

Car ferry in Split, Croatia

Other Modes

The Trans-Alaska Pipeline System conveys crude oil from the Prudhoe Bay Oil Field to Valdez, Alaska

Pipeline transport sends goods through a pipe, most commonly liquid and gases are sent, but pneumatic tubes can also send solid capsules using compressed air. For example liquids/gases, any chemically stable liquid or gas can be sent through a pipeline. Short-distance systems exist for sewage, slurry water and beer, while long-distance networks are used for petroleum and natural gas.

Cable transport is a broad mode where vehicles are pulled by cables instead of an internal power source. It is most commonly used at steep gradient. Typical solutions include aerial tramway, elevators, escalator and ski lifts; some of these are also categorized as conveyor transport.

Space transport is transport out of Earth's atmosphere into outer space by means of a spacecraft. While large amounts of research have gone into technology, it is rarely used except to put satellites into orbit, and conduct scientific experiments. However, man has landed on the moon, and probes have been sent to all the planets of the Solar System.

Unmanned aerial vehicle transport (drone transport) is currently being tested by Amazon.com and other transportation companies. This method will allow short-range small-parcel delivery in a short time frame.

Components of a Mode of Transport

A transport mode is a combination of the following:

- Transportation infrastructure: thoroughfares, networks, hubs (stations, bus terminals, airport terminals), etc.
- Vehicles and containers: motor vehicles, automobiles, motorcycles, trucks, wagons, trains, ships, and aircraft
- A stationary or mobile workforce
- Propulsion system and power supply (traction)
- Operations: driving, management, traffic signals, railway signalling, air traffic control, etc.

Purpose of Transport : Freight & Passenger movement and mobility are core components of a transport system.

Worldwide Comparison of The Most Important Transport

Worldwide, the most widely used modes for passenger transport are the Automobile (16,000 bn passenger km), followed by Buses (7,000), Air (2,800), Railways (1,900), and Urban Rail (250).

The most widely used modes for freight transport are Sea (40,000 bn ton km), followed by Road (7,000), Railways (6,500), Oil pipelines (2,000) and Inland Navigation (1,500).

	EU 15	**USA**	**Japan**	**World**
GDP (PPP) per capita (€)	19,000	28,600	22,300	5,500
Passenger km per capita				
Private car	10,100	22,700	6,200	2,700
Bus/ coach	1,050	870	740	1,200
Railway	750	78	2,900	320
Air (domestic except World)	860	2,800	580	480

Transport Network

Transportation networks generally refer to a set of links, nodes, and lines that represent the infrastructure or supply side of the transportation. The links have characteristics such as speed and capacity for roadways; frequency and travel time data are defined on transit links or lines for the transit system. Transportation analysis zones are represented in the network by centroids, a subset of the nodes; and centroid connectors are links that provide access and egress to the system.

Highway and transit networks are a principal means by which the supply side of transportation is represented. A typical highway network will include links ranging from high speed, high capacity (e.g., freeways) to low-speed, low capacity (e.g., residential streets). Transit networks are a spatial representation of bus, rail, and other types of transit routes available in a region. Links will include fixed-guideway (e.g., passenger rail) to bus lines that operate in mixed-traffic on the roadway links.

While all models are representations of the 'real world', it should be noted that network models can be both very precise and accurate. The following quote puts this in perspective especially as it relates to the differences between the supply and demand side of models.

> "Traffic modeling is frequently described as being part science, part art. Modelling the base year supply-side -- the representation of the highway network today -- is the science part. It is frequently modelled with, literally, military precision using mapping data from GPS satelittes. This level of precision does not extend to the representation of the demand-side of the travel economy..."

Network Types

Highway Networks

Highway networks provide a representation of individual, connected links between intersections through the use of *links* and *nodes*. Links are often described by length, capacity, number of lanes, posted speed, and freeflow speed. The congested speed on any link may be calculated through the use of an appropriate volume-delay function (VDF).

Transit Networks

Transit networks are also comprised of nodes and links. Nodes generally comprise bus or rail stops along transit routes and are spatially located by XY coordinates. Stop attributes may be attached to a node to describe the characteristics of the stop, typical dwell times, and frequency of vehicle arrivals at the stop. Links on the transit network connect consecutive stops. Further detail about each route often includes route fares, schedules, headways, and speeds. Other spatial data elements associated with transit networks include the locations and capacity of park-n-ride facilities, and access and egress legs or links for each zone. Intra- and intermodal transfer locations are also identified in transit networks.

Other Networks

Non-motorized modes such as walking or bicycling are sometimes not represented in travel forecasting models. Hence, the networks for these modes would not included. The main reason for this exclusion is that most models were developed to address macro-level questions about the location or size of the roadway and transit network links (i.e., are new roads needed or what existing roads need to be widened). However, an emerging area of interest includes bike modeling, which requires a network that may include special attributes that influence route choice (such as street slope).

Freight Networks can Include both Truck Facilities and Rail Freight.

Super networks represent an integrated multi-modal network with the intention of providing a more comprehensive look at routes across all available modes of transport and access/egress patterns. The modal elements appear in multiple layers and are connected by artificial transfer links, which can represent a variety of costs. The private (car and bicycle) layers can be connected to the pedestrian layer where they are available and can be parked, and all origin and destination centroids are located on the pedestrian layer.

Methods

Transport network analysis is used to determine the flow of vehicles (or people) through a transport network, typically using mathematical graph theory. It may combine different modes of transport, for example, walking and car, to model multi-modal journeys. Transport network analysis falls within the field of transport engineering. Traffic has been studied extensively using statistical physics methods.

Recently a real transport network of Beijing was studied using network approach and percolation theory.

Land Transport

Providing and managing the land transport system is a key issue facing regional and district councils as well as other providers of land transport. There is a need for strong relationships and integration between policies and strategies at both a strategic and plan level.

'Land transport' covers all land-based transportation systems that provide for the movement of people, goods and services, and includes the following:

- road networks from the state highways to local roads
- rail networks
- provisions for pedestrians and cyclists
- public transport networks (services and infrastructure).

The land transport system plays a vital role in linking communities within a district and linking those communities with other districts and regions. As a physical resource of strategic importance, the land transport system needs to be managed, and protected from adverse effects (that may arise from land uses) that could affect the provision of an integrated, safe, responsive and sustainable land transport. Similarly, adverse environmental effects on land use activities resulting from land transport systems also need to be managed.

Specific concerns in relation to land transport include:

- increased reliance on single-occupant motor vehicle trips, and increased vehicle kilometres
- declining future car use for certain groups (e.g. younger drivers) and the need for flexibility to enable the system to respond
- growing congestion on the road network, and increased journey times

- reduced ability for travel needs to be met by public transport and walking and cycling
- reduced accessibility for older people and others in society who do not have a car or are unable to use one
- ribbon and other dispersed development along arterial routes, with vehicles moving into and out of these developments, impeding the free flow of traffic, increasing the probability of crashes, and creating pressures for bypasses
- incompatible land use and transport development (e.g. residents in new housing estates built alongside arterial routes opposing further development of the routes because of concerns about traffic noise and other environmental impacts).

Systems

Several systems of land transport have been devised, from the most basic system of humans carrying things from place to sophisticated networks of ground-based transportation utilizing different types of vehicles and infrastructure. The three types are human-powered, animal powered and machine powered

Human-powered Transportation

Human-powered transport remains common in developing countries.

Human powered transport, a form of sustainable transportation, is the transport of people and/or goods using human muscle-power, in the form of walking, running and swimming. Modern technology has allowed machines to enhance human power. Human-powered transport remains popular for reasons of cost-saving, leisure, physical exercise, and environmentalism; it is sometimes the only type available, especially in underdeveloped or inaccessible regions.

Although humans are able to walk without infrastructure, the transport can be enhanced through the use of roads, especially when using the human power with vehicles, such as bicycles and inline skates. Human-powered vehicles have also been developed for difficult environments, such as snow and water, by watercraft rowing and skiing; even the air can be entered with human-powered aircraft.

Animal-powered Transportation

Animal-powered transport is the use of working animals for the movement of people and goods. Humans may ride some of the animals directly, use them as pack animals for carrying goods, or harness them, alone or in teams, to pull sleds or wheeled vehicles.

Road Transportation

The Harbor Freeway is often heavily congested at rush hour in Downtown Los Angeles.

A road is an identifiable route, way or path between two or more places. Roads are typically smoothed, paved, or otherwise prepared to allow easy travel; though they need not be, and historically many roads were simply recognizable routes without any formal construction or maintenance. In urban areas, roads may pass through a city or village and be named as streets, serving a dual function as urban space easement and route.

The most common road vehicle is the automobile; a wheeled passenger vehicle that carries its own motor. Other users of roads include buses, trucks, motorcycles, bicycles and pedestrians. As of 2002, there were 590 million automobiles worldwide. Automobiles have caused air pollution which results to people getting Asthma . Road transport offers a complete freedom to road users to transfer the vehicle from one lane to the other and from one road to another according to the need and convenience. This flexibility of changes in location, direction, speed, and timings of travel is not available to other modes of transport. It is possible to provide door to door service only by road transport.

Automobiles offer high flexibility and with low capacity, but are deemed with high energy and area use, and the main source of noise and air pollution in cities; buses allow for more efficient travel at the cost of reduced flexibility. Road transport by truck is often the initial and final stage of freight transport.

Rail Transportation

InterCityExpress, a German high-speed passenger train

Rail transport is where a train runs along a set of two parallel steel rails, known as a railway or railroad. The rails are anchored perpendicular to ties (or sleepers) of timber, concrete or steel, to maintain a consistent distance apart, or gauge. The rails and perpendicular beams are placed on a foundation made of concrete, or compressed earth and gravel in a bed of ballast. Alternative methods include monorail and maglev.

A train consists of one or more connected vehicles that run on the rails. Propulsion is commonly provided by a locomotive, that hauls a series of unpowered cars, that can carry passengers or freight. The locomotive can be powered by steam, diesel or by electricity supplied by trackside systems. Alternatively, some or all the cars can be powered, known as a multiple unit. Also, a train can be powered by horses, cables, gravity, pneumatics and gas turbines. Railed vehicles move with much less friction than rubber tires on paved roads, making trains more energy efficient, though not as efficient as ships.

Intercity trains are long-haul services connecting cities; modern high-speed rail is capable of speeds up to 350 km/h (220 mph), but this requires specially built track. Regional and commuter trains feed cities from suburbs and surrounding areas, while intra-urban transport is performed by high-capacity tramways and rapid transits, often making up the backbone of a city's public transport. Freight trains traditionally used box cars, requiring manual loading and unloading of the cargo. Since the 1960s, container trains have become the dominant solution for general freight, while large quantities of bulk are transported by dedicated trains.

Other Modes

Pipeline transport sends goods through a pipe; most commonly liquid and gases are sent, but pneumatic tubes can also send solid capsules using compressed air. For liquids/gases, any chemically stable liquid or gas can be sent through a pipeline. Short-distance systems exist for sewage, slurry, water and beer, while long-distance networks are used for petroleum and natural gas.

Cable transport is a broad mode where vehicles are pulled by cables instead of an internal power source. It is most commonly used at steep gradient. Typical solutions include aerial tramway, elevators, escalator and ski lifts; some of these are also categorized as conveyor transport.

Trans-Alaska Pipeline for crude oil

Connections with other Modes

Airports

Airports serve as a terminus for air transport activities, but most people and cargo transported by air must use ground transport to reach their final destination.

Airport-based services are sometimes used to shuttle people to nearby hotels or motels when overnight stay is required for connecting flights. Companies provide rental car, private bus and taxi services while mass transportation is usually provided by a municipality or other source of public funding.

Several major airports, including Denver International and JFK International, provide many types of ground transportation, often by working with livery companies and similar businesses. Smaller airports might only have a few private rental companies and a bus service. Larger airports tend to offer several different transportation options. Larger airports also sometimes have light rail and/ or roads that loop around an airport to provide access from multiple terminals.

Seaports

As with air transport, sea transport typically requires use of ground transport at either end of travel for people and goods to reach their final destinations. Significant infrastructure is used at ports to transfer people and goods between sea and land systems.

Elements

Infrastructure

Infrastructure is the fixed installations that allow a vehicle to operate. It consists of a way, a terminal and facilities for parking and maintenance. For rail, pipeline, road and cable transport, the entire way the vehicle travels must be built up.

Terminals such as stations are locations where passengers and freight can be transferred from one vehicle or mode to another. For passenger transport, terminals are integrating different modes to allow riders to interchange to take advantage of each mode's advantages. For instance, airport rail links connect airports to the city centers and suburbs. The terminals for automobiles are parking lots, while buses and coaches can operate from simple stops. For freight, terminals act as transshipment points, though some cargo is transported directly from the point of production to the point of use.

The financing of infrastructure can either be public or private. Transport is often a natural monopoly and a necessity for the public; roads, and in some countries railways and airports are funded through taxation. New infrastructure projects can have high cost, and are often financed through debt. Many infrastructure owners therefore impose usage fees, such as landing fees at airports, or toll plazas on roads. Independent of this, authorities may impose taxes on the purchase or use of vehicles. Because of poor forecasting and overestimation of passenger numbers by planners, there is frequently a benefits shortfall for transport infrastructure projects.

Bridges, such as Golden Gate Bridge, allow roads and railways to cross bodies of water.

Vehicles

A Fiat Uno

A vehicle is any non-living device that is used to move people and goods. Unlike the infrastructure, the vehicle moves along with the cargo and riders. Unless being pulled by a cable or muscle-power, the vehicle must provide its own propulsion; this is most commonly done through a steam engine, combustion engine, or electric motor, though other means of propulsion also exist. Vehicles also need a system of converting the energy into movement; this is most commonly done through wheels, propellers and pressure.

Vehicles are most commonly staffed by a driver. However, some systems, such as people movers and some rapid transits, are fully automated. For passenger transport, the vehicle must have a compartment for the passengers. Simple vehicles, such as automobiles, bicycles or simple aircraft, may have one of the passengers as a driver.

Users

Public

Public land transport refers to carriage of people and goods by government or commercial entities which is made available to the public at large for the purpose of facilitating the economy and society they serve. Most transport infrastructure and large transport vehicles are operated in this manner. Funds to pay for such transport may come from taxes, subscriptions, direct user fees, or some combination. The vast majority of public transport is land-based, with commuting and postal delivery being the primary purposes.

Commerce

Commercial land transport refers to carriage of people and goods by commercial entities made available at cost to individuals, businesses, and the government for the purpose of profiting the entities providing the travel. Most infrastructure used is publicly owned, and vehicles tend to be large and efficient to maximize capacity and profit margins. Freight shipping and long-distance travel are common uses served by commercial land transport.

Military

Military land transport refers to carriage of people and goods by the military or other operators for the purpose of supporting military operations, both in peacetime as well as in combat areas. Such activity may use a combination of public infrastructure as well as military-specific infrastructure and in many cases is designed to operate with little or no infrastructure when necessary. Vehicles can range from basic commercial or even private vehicles to those specifically designed for military use.

Private

Private land transport refers to individuals and organizations transporting themselves and their own people, animals, and goods at their own discretion. Vehicles used are typically smaller, though publicly owned infrastructure is often used for travel.

Function

Relocation of travelers and cargo are the most common uses of transport. However, other uses exist, such as the strategic and tactical relocation of armed forces during warfare, or the civilian mobility construction or emergency equipment.

Passenger

A local transit bus operated by ACTION in Canberra, Australia

Passenger transport, or travel, is divided into public and private transport. Public transport is scheduled services on fixed routes, while private is vehicles that provide ad hoc services at the riders desire. The latter offers better flexibility, but has lower capacity, and a higher environmental impact. Travel may be as part of daily commuting, for business, leisure or migration.

Short-haul transport is dominated by the automobile and mass transit. The latter consists of buses in rural and small cities, supplemented with commuter rail, trams and rapid transit in larger cities. Long-haul transport involves the use of the automobile, trains, coaches and aircraft, the last of which have become predominantly used for the longest, including intercontinental, travel. Intermodal passenger transport is where a journey is performed through the use of several modes of transport; since all human transport normally starts and ends with walking, all passenger transport can be considered intermodal. Public transport may also involve the intermediate change of vehicle, within or across modes, at a transport hub, such as a bus or railway station.

Taxis and buses can be found on both ends of the public transport spectrum. Buses are the cheaper mode of transport but are not necessarily flexible, and taxis are very flexible but more expensive. In the middle is demand-responsive transport, offering flexibility whilst remaining affordable.

International travel may be restricted for some individuals due to legislation and visa requirements.

Freight

Freight transport, or shipping, is a key in the value chain in manufacturing. With increased specialization and globalization, production is being located further away from consumption, rapidly increasing the demand for transport. While all modes of transport are used for cargo transport, there is high differentiation between the nature of the cargo transport, in which mode is chosen. Logistics refers to the entire process of transferring products from producer to consumer, including storage, transport, transshipment, warehousing, material-handling and packaging, with associated exchange of information. Incoterm deals with the handling of payment and responsibility of risk during transport.

Freight train with shipping containers in the United Kingdom

Containerization, with the standardization of ISO containers on all vehicles and at all ports, has revolutionized international and domestic trade, offering huge reduction in transshipment costs. Traditionally, all cargo had to be manually loaded and unloaded into the haul of any car; containerization allows for automated handling and transfer between modes, and the standardized sizes allow for gains in economy of scale in vehicle operation. This has been one of the key driving factors in international trade and globalization since the 1950s.

Bulk transport is common with cargo that can be handled roughly without deterioration; typical examples are ore, coal, cereals and petroleum. Because of the uniformity of the product, mechanical handling can allow enormous quantities to be handled quickly and efficiently. The low value of the cargo combined with high volume also means that economies of scale become essential in transport, and whole trains are commonly used to transport bulk. Liquid products with sufficient volume may also be transported by pipeline.

Rail Transport

Rail transport is also known as train transport. It is a means of transport, on vehicles which run on tracks (rails or railroads). It is one of the most important, commonly used and very cost effective modes of commuting and goods carriage over long, as well as, short distances. Since this system runs on metal (usually steel) rails and wheels, it has an inherent benefit of lesser frictional resistance which helps attach more load in terms of wagons or carriages. This system is known as a train. Usually, trains are powered by an engine locomotive running on electricity or on diesel. Complex signaling systems are utilised if there are multiple route networks. Rail transport is also one of the fastest modes of land transport.

Rail transport is one of the oldest modes of transportation that is suitable for conveying bulky goods over long distances. Railway is perhaps a sobering thought that until the railways emerged in the 1830s man could travel no faster than a horse could gallop. These appeared during Brittanâ€™s industrial Revolution and later were constructed throughout the developed world.

However, railways were developed during the period of industrial revolution in the 19th century, these was partly for political reasons and for economic reasons. In many countries, they were built especially to penetrate isolated regions and help promote political unity. The major advantage of railway transport includes provision reliable services. Â For instance, Russia was crossed be the Tans- Siberian Railway; North America was crossed by the Canadian Pacific Railway. In other places in the world, railways aide the exploitation of mineral deposits as in the case of Australia and Latin America. In Britain, railways where the most important land transport system throughout the Victorian age. Only fairly recently have they declined in significance as a result of an increase in the use of motor vehicles and on the hand it is as a result of greater air traffic in Britain.

Description

Rail transport has emerged as one of the most dependable modes of transport in terms of safety. Trains are fast and the least affected by usual weather turbulences like rain or fog, compared to other transport mechanisms. Rail transport is better organised than any other medium of transport. It has fixed routes and schedules. Its services are more certain, uniform and regular compared to other modes of transport. Rail transport originated from human hauled contraptions in ancient Greece. Now it has evolved into a modern, complex and sophisticated system used both in urban and cross-country (and continent) networks over long distances.

Rail transport is an enabler of economic progress, used to mobilise goods as well as people. Adaptations include passenger railways, underground (or over ground) urban metro railways and goods carriages. Rail transport has some constraints and limitations also. One of the biggest constraints of rail transport is heavy cost. Trains need high capital to build and maintain and the cost is magnified when a whole rail network is to be built. The cost of construction, maintenance and overhead expenses are very high compared to other modes of transport. Also, rail transport cannot provide door-to-door service as it is tied to a particular track. Intermediate loading or unloading involves greater cost, more wear and tear and wastage of time.

Trains

A train is a connected series of rail vehicles that move along the track. Propulsion for the train is provided by a separate locomotive or from individual motors in self-propelled multiple units. Most trains carry a revenue load, although non-revenue cars exist for the railway's own use, such as for maintenance-of-way purposes. The engine driver (engineer in North America) controls the locomotive or other power cars, although people movers and some rapid transits are under automatic control.

Haulage

Traditionally, trains are pulled using a locomotive. This involves one or more powered vehicles being located at the front of the train, providing sufficient tractive force to haul the weight of the full train. This arrangement remains dominant for freight trains and is often used for passenger trains. A push-pull train has the end passenger car equipped with a driver's cab so that the engine driver can remotely control the locomotive. This allows one of the locomotive-hauled train's drawbacks to be removed, since the locomotive need not be moved to the front of the train each time the train changes direction. A railroad car is a vehicle used for the haulage of either passengers or freight.

A multiple unit has powered wheels throughout the whole train. These are used for rapid transit and tram systems, as well as many both short- and long-haul passenger trains. A railcar is a single, self-powered car, and may be electrically-propelled or powered by a diesel engine. Multiple units have a driver's cab at each end of the unit, and were developed following the ability to build electric motors and engines small enough to fit under the coach. There are only a few freight multiple units, most of which are high-speed post trains.

Russian 2TE10U Diesel-electric locomotive

Motive Power

Steam locomotives are locomotives with a steam engine that provides adhesion. Coal, petroleum, or wood is burned in a firebox, boiling water in the boiler to create pressurized steam. The steam travels through the smokebox before leaving via the chimney or smoke stack. In the process, it powers a piston that transmits power directly through a connecting rod (US: main rod) and a crankpin (US: wristpin) on the driving wheel (US main driver) or to a crank on a driving axle. Steam locomotives have been phased out in most parts of the world for economical and safety reasons, although many are preserved in working order by heritage railways.

Electric locomotives draw power from a stationary source via an overhead wire or third rail. Some also or instead use a battery. In locomotives that are powered by high voltage alternating current, a transformer in the locomotive converts the high voltage, low current power to low voltage, high current used in the traction motors that power the wheels. Modern locomotives may use three-phase AC induction motors or direct current motors. Under certain conditions, electric locomotives are the most powerful traction. They are also the cheapest to run and provide less noise and no local air pollution. However, they require high capital investments both for the overhead lines and the supporting infrastructure, as well as the generating station that is needed to produce electricity. Accordingly, electric traction is used on urban systems, lines with high traffic and for high-speed rail.

Diesel locomotives use a diesel engine as the prime mover. The energy transmission may be either diesel-electric, diesel-mechanical or diesel-hydraulic but diesel-electric is dominant. Electro-diesel locomotives are built to run as diesel-electric on unelectrified sections and as electric locomotives on electrified sections.

Alternative methods of motive power include magnetic levitation, horse-drawn, cable, gravity, pneumatics and gas turbine.

A RegioSwinger multiple unit of the Croatian Railways

Passenger trains

Interior view of the top deck of a VR InterCity2 double-deck carriage

A passenger train travels between stations where passengers may embark and disembark. The oversight of the train is the duty of a guard/train manager/conductor. Passenger trains are part of public transport and often make up the stem of the service, with buses feeding to stations. Passenger trains provide long-distance intercity travel, daily commuter trips, or local urban transit services. They even include a diversity of vehicles, operating speeds, right-of-way requirements, and service frequency. Passenger trains usually can be divided into two operations: intercity railway and intracity transit. Whereas as intercity railway involve higher speeds, longer routes, and lower frequency (usually scheduled), intracity transit involves lower speeds, shorter routes, and higher frequency (especially during peak hours).

Intercity trains are long-haul trains that operate with few stops between cities. Trains typically have amenities such as a dining car. Some lines also provide over-night services with sleeping cars. Some long-haul trains have been given a specific name. Regional trains are medium distance trains that connect cities with outlying, surrounding areas, or provide a regional service, making more stops and having lower speeds. Commuter trains serve suburbs of urban areas, providing a daily commuting service. Airport rail links provide quick access from city centres to airports.

High-speed rail are special inter-city trains that operate at much higher speeds than conventional railways, the limit being regarded at 200 to 320 kilometres per hour (120 to 200 mph). High-speed trains are used mostly for long-haul service and most systems are in Western Europe and East Asia. The speed record is 574.8 km/h (357.2 mph), set by a modified French TGV. Magnetic levitation trains such as the Shanghai airport train use under-riding magnets which attract themselves upward towards the underside of a guideway and this line has achieved somewhat higher peak speeds in day-to-day operation than conventional high-speed railways, although only over short distances. Due to their heightened speeds, route alignments for high-speed rail tend to have shallower grades and broader curves than conventional railways.

Their high kinetic energy translates to higher horsepower-to-ton ratios (e.g. 20 horsepower per short ton or 16 kilowatts per tonne); this allows trains to accelerate and maintain higher speeds and negotiate steep grades as momentum builds up and recovered in downgrades (reducing cut, fill, and tunnelling requirements). Since lateral forces act on curves, curvatures are designed with the highest possible radius. All these features are dramatically different from freight operations, thus justifying exclusive high-speed rail lines if it is economically feasible.

Rail network in Paris, France

Higher-speed rail services are intercity rail services that have top speeds higher than conventional intercity trains but the speeds are not as high as those in the high-speed rail services. These services are provided after improvements to the conventional rail infrastructure in order to support trains that can operate safely at higher speeds.

Rapid transit is an intracity system built in large cities and has the highest capacity of any passenger transport system. It is usually grade-separated and commonly built underground or elevated.

At street level, smaller trams can be used. Light rails are upgraded trams that have step-free access, their own right-of-way and sometimes sections underground. Monorail systems are elevated, medium-capacity systems. A people mover is a driverless, grade-separated train that serves only a few stations, as a shuttle. Due to the lack of uniformity of rapid transit systems, route alignment varies, with diverse rights-of-way (private land, side of road, street median) and geometric characteristics (sharp or broad curves, steep or gentle grades). For instance, the Chicago 'L' trains are designed with extremely short cars to negotiate the sharp curves in the Loop. New Jersey's PATH has similar-sized cars to accommodate curves in the trans-Hudson tunnels. San Francisco's BART operates large cars on its well-engineered routes.

Freight Train

Bulk cargo of minerals

A freight train hauls cargo using freight cars specialized for the type of goods. Freight trains are very efficient, with economy of scale and high energy efficiency. However, their use can be reduced by lack of flexibility, if there is need of transshipment at both ends of the trip due to lack of tracks to the points of pick-up and delivery. Authorities often encourage the use of cargo rail transport due to its fame.

Container trains have become the beta type in the US for bulk haulage. Containers can easily be transshipped to other modes, such as ships and trucks, using cranes. This has succeeded the boxcar (wagon-load), where the cargo had to be loaded and unloaded into the train manually. The intermodal containerization of cargo has revolutionized the supply chain logistics industry, reducing ship costs significantly. In Europe, the sliding wall wagon has largely superseded the ordinary covered wagons. Other types of cars include refrigerator cars, stock cars for livestock and autoracks for road vehicles. When rail is combined with road transport, a roadrailer will allow trailers to be driven onto the train, allowing for easy transition between road and rail.

Bulk handling represents a key advantage for rail transport. Low or even zero transshipment costs combined with energy efficiency and low inventory costs allow trains to handle bulk much cheaper than by road. Typical bulk cargo includes coal, ore, grains and liquids. Bulk is transported in open-topped cars, hopper cars and tank cars.

Infrastructure

Above: Railway turnouts; Below: Chicago Transit Authority control tower 18 guides elevated Chicago 'L' north and southbound Purple and Brown lines intersecting with east and westbound Pink and Green lines and the looping Orange line above the Wells and Lake street intersection in the loop at an elevated right of way.

Right of Way

Railway tracks are laid upon land owned or leased by the railway company. Owing to the desirability of maintaining modest grades, rails will often be laid in circuitous routes in hilly or mountainous terrain. Route length and grade requirements can be reduced by the use of alternating cuttings, bridges and tunnels – all of which can greatly increase the capital expenditures required to develop a right of way, while significantly reducing operating costs and allowing higher speeds on longer radius curves. In densely urbanized areas, railways are sometimes laid in tunnels to minimize the effects on existing properties.

Trackage

Track consists of two parallel steel rails, anchored perpendicular to members called ties (sleepers) of timber, concrete, steel, or plastic to maintain a consistent distance apart, or rail gauge. Rail gauges are usually categorized as standard gauge (used on approximately 54.8% of the world's existing railway lines), broad gauge, and narrow gauge. In addition to the rail gauge, the tracks will be laid to conform with a Loading gauge which defines the maximum height and width for railway vehicles and their loads to ensure safe passage through bridges, tunnels and other structures.

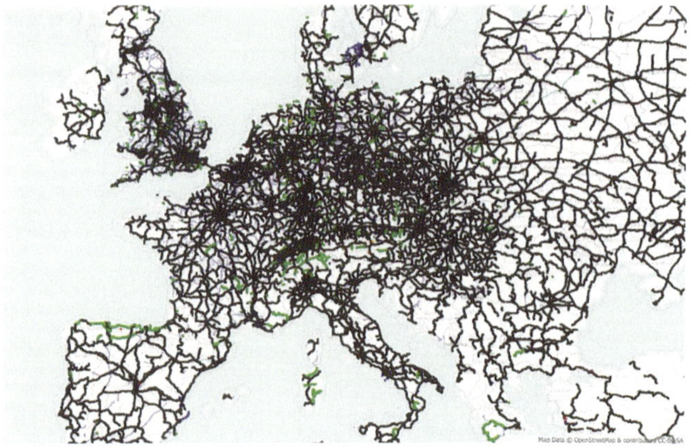

Map of railways in Europe with main operational lines shown in black, heritage railway lines in green and former routes in light blue

Long freight train crossing the Stoney Creek viaduct on the Canadian Pacific Railway in southern British Columbia

The track guides the conical, flanged wheels, keeping the cars on the track without active steering and therefore allowing trains to be much longer than road vehicles. The rails and ties are usually placed on a foundation made of compressed earth on top of which is placed a bed of ballast to distribute the load from the ties and to prevent the track from buckling as the ground settles over time under the weight of the vehicles passing above.

The ballast also serves as a means of drainage. Some more modern track in special areas is attached by direct fixation without ballast. Track may be prefabricated or assembled in place. By welding rails together to form lengths of continuous welded rail, additional wear and tear on rolling stock caused by the small surface gap at the joints between rails can be counteracted; this also makes for a quieter ride (passenger trains).

On curves the outer rail may be at a higher level than the inner rail. This is called superelevation or cant. This reduces the forces tending to displace the track and makes for a more comfortable ride for standing livestock and standing or seated passengers. A given amount of superelevation is most effective over a limited range of speeds.

Turnouts, also known as points and switches, are the means of directing a train onto a diverging section of track. Laid similar to normal track, a point typically consists of a frog (common crossing), check rails and two switch rails. The switch rails may be moved left or right, under the control of the signalling system, to determine which path the train will follow.

Spikes in wooden ties can loosen over time, but split and rotten ties may be individually replaced with new wooden ties or concrete substitutes. Concrete ties can also develop cracks or splits, and can also be replaced individually. Should the rails settle due to soil subsidence, they can be lifted by specialized machinery and additional ballast tamped under the ties to level the rails.

Periodically, ballast must be removed and replaced with clean ballast to ensure adequate drainage. Culverts and other passages for water must be kept clear lest water is impounded by the trackbed, causing landslips. Where trackbeds are placed along rivers, additional protection is usually placed to prevent streambank erosion during times of high water. Bridges require inspection and maintenance, since they are subject to large surges of stress in a short period of time when a heavy train crosses.

Train Inspection Systems

A Hot bearing detector w/ dragging equipment unit

The inspection of railway equipment is essential for the safe movement of trains. Many types of defect detectors are in use on the world's railroads. These devices utilize technologies that vary from a simplistic paddle and switch to infrared and laser scanning, and even ultrasonic audio analysis. Their use has avoided many rail accidents over the 70 years they have been used.

Signalling

Bardon Hill box in England is a Midland Railway box, although the original mechanical lever frame has been replaced by electrical switches

Railway signalling is a system used to control railway traffic safely to prevent trains from colliding. Being guided by fixed rails which generate low friction, trains are uniquely susceptible to collision since they frequently operate at speeds that do not enable them to stop quickly or within the driver's sighting distance; road vehicles, which encounter a higher level of friction between their rubber tyres and the road surface, have much shorter braking distances. Most forms of train control involve movement authority being passed from those responsible for each section of a rail network to the train crew. Not all methods require the use of signals, and some systems are specific to single track railways.

The signalling process is traditionally carried out in a signal box, a small building that houses the lever frame required for the signalman to operate switches and signal equipment. These are placed at various intervals along the route of a railway, controlling specified sections of track. More recent technological developments have made such operational doctrine superfluous, with the centralization of signalling operations to regional control rooms. This has been facilitated by the increased use of computers, allowing vast sections of track to be monitored from a single location. The common method of block signalling divides the track into zones guarded by combinations of block signals, operating rules, and automatic-control devices so that only one train may be in a block at any time.

Electrification

The electrification system provides electrical energy to the trains, so they can operate without a prime mover on board. This allows lower operating costs, but requires large capital investments along the lines. Mainline and tram systems normally have overhead wires, which hang from poles along the line. Grade-separated rapid transit sometimes use a ground third rail.

Power may be fed as direct or alternating current. The most common DC voltages are 600 and 750 V for tram and rapid transit systems, and 1,500 and 3,000 V for mainlines. The two dominant AC systems are 15 kV AC and 25 kV AC.

Stations

Goods station in Lucerne, Switzerland

A railway station serves as an area where passengers can board and alight from trains. A goods station is a yard which is exclusively used for loading and unloading cargo. Large passenger stations have at least one building providing conveniences for passengers, such as purchasing tickets and food. Smaller stations typically only consist of a platform. Early stations were sometimes built with both passenger and goods facilities.

Platforms are used to allow easy access to the trains, and are connected to each other via underpasses, footbridges and level crossings. Some large stations are built as culs-de-sac, with trains only operating out from one direction. Smaller stations normally serve local residential areas, and may have connection to feeder bus services. Large stations, in particular central stations, serve as the main public transport hub for the city, and have transfer available between rail services, and to rapid transit, tram or bus services.

Operations

Ownership

In the United States, railroads such as the Union Pacific traditionally own and operate both their rolling stock and infrastructure, with the company itself typically being privately owned.

Since the 1980s, there has been an increasing trend to split up railway companies, with companies owning the rolling stock separated from those owning the infrastructure. This is particularly true in Europe, where this arrangement is required by the European Union. This has allowed open access by any train operator to any portion of the European railway network. In the UK, the railway track is state owned, with a public controlled body (Network Rail) running, maintaining and developing the track, while Train Operating Companies have run the trains since privatization in the 1990s.

In the U.S., virtually all rail networks and infrastructure outside the Northeast Corridor are privately owned by freight lines. Passenger lines, primarily Amtrak, operate as tenants on the freight lines. Consequently, operations must be closely synchronized and coordinated between freight and passenger railroads, with passenger trains often being dispatched by the host freight railroad. Due to this shared system, both are regulated by the Federal Railroad Administration (FRA) and may follow the AREMA recommended practices for track work and AAR standards for vehicles.

Financing

The main source of income for railway companies is from ticket revenue (for passenger transport) and shipment fees for cargo. Discounts and monthly passes are sometimes available for frequent travellers (e.g. season ticket and rail pass). Freight revenue may be sold per container slot or for a whole train. Sometimes, the shipper owns the cars and only rents the haulage. For passenger transport, advertisement income can be significant.

Governments may choose to give subsidies to rail operation, since rail transport has fewer externalities than other dominant modes of transport. If the railway company is state-owned, the state may simply provide direct subsidies in exchange for increased production. If operations have been privatized, several options are available. Some countries have a system where the infrastructure is owned by a government agency or company – with open access to the tracks for any company that meets safety requirements. In such cases, the state may choose to provide the tracks free of charge, or for a fee that does not cover all costs. This is seen as analogous to the government providing free access to roads. For passenger operations, a direct subsidy may be paid to a public-owned operator, or public service obligation tender may be helt, and a time-limited contract awarded to the lowest bidder. Total EU rail subsidies amounted to €73 billion in 2005.

Amtrak, the US passenger rail service, and Canada's Via Rail are private railroad companies chartered by their respective national governments. As private passenger services declined because of competition from automobiles and airlines, they became shareholders of Amtrak either with a cash entrance fee or relinquishing their locomotives and rolling stock. The government subsidizes Amtrak by supplying start-up capital and making up for losses at the end of the fiscal year.

Safety

Trains can travel at very high speed, but they are heavy, are unable to deviate from the track and require a great distance to stop. Possible accidents include derailment (jumping the track), a collision with another train or collision with automobiles, other vehicles or pedestrians at level crossings. The last accounts for the majority of rail accidents and casualties. The most important safety measures to prevent accidents are strict operating rules, e.g. railway signalling and gates or grade separation at crossings. Train whistles, bells or horns warn of the presence of a train, while trackside signals maintain the distances between trains.

An important element in the safety of many high-speed inter-city networks such as Japan's Shinkansen is the fact that trains only run on dedicated railway lines, without level crossings. This effectively eliminates the potential for collision with automobiles, other vehicles or pedestrians, vastly reduces the likelihood of collision with other trains and helps ensure services remain timely.

Fatality risk of passenger per mode of transport in European Union	
Transport mode used by user	Fatalities per billion passenger kilometers
Airline passenger	0.101
Railway passenger	0.156
Bus/Coach occupant	0.433
Car occupant	4.450
Powered two-wheelers	52.593

Fatality risk ratios for transports	Airline passenger	Railway passenger	Bus/Coach occupant	Car occupant	Powered two-wheelers
Powered two-wheelers	520	337	121	12	1
Car occupant	44	28.5	10	1	
Bus/Coach occupant	4.3	2.8	1		
Railway passenger	1.5	1			
Airline passenger	1				

According to Eurostat and European Railway Agency, on European railways, there is a fatality risk for passengers and occupants 28 times lower compared with car usage. Based on data by EU-27 member nations

Maintenance

As in any infrastructure asset, railways must keep up with periodic inspection and maintenance in order to minimize effect of infrastructure failures that can disrupt freight revenue operations and passenger services. Because passengers are considered the most *crucial cargo* and usually operate at higher speeds, steeper grades, and higher capacity/frequency, their lines are especially important. Inspection practices include track geometry cars or walking inspection. Curve maintenance especially for transit services includes gauging, fastener tightening, and rail replacement.

Rail corrugation is a common issue with transit systems due to the high number of light-axle, wheel passages which result in grinding of the wheel/rail interface. Since maintenance may overlap with operations, maintenance windows (nighttime hours, off-peak hours, altering train schedules or routes) must be closely followed. In addition, passenger safety during maintenance work (inter-track fencing, proper storage of materials, track work notices, hazards of equipment near states) must be regarded at all times. At times, maintenance access problems can emerge due to tunnels, elevated structures, and congested cityscapes. Here, specialized equipment or smaller versions of conventional maintenance gear are used.

Unlike highways or road networks where capacity is disaggregated into unlinked trips over individual route segments, railway capacity is fundamentally considered a network system. As a result, many components are causes and effects of system disruptions. Maintenance must acknowledge the vast array of a route's performance (type of train service, origination/destination, seasonal impacts), line's capacity (length, terrain, number of tracks, types of train control), trains throughput (max speeds, acceleration/deceleration rates), and service features with shared passenger-freight tracks (sidings, terminal capacities, switching routes, and design type).

Social, Economical, and Energetic Aspects

Energy

BNSF Railway freight service in the United States

German InterCityExpress (ICE)

Rail transport is an energy-efficient but capital-intensive means of mechanized land transport. The tracks provide smooth and hard surfaces on which the wheels of the train can roll with a relatively low level of friction being generated. Moving a vehicle on and/or through a medium (land, sea, or air) requires that it overcomes resistance to its motion caused by friction. A land vehicle's total resistance (in pounds or Newtons) is a quadratic function of the vehicle's speed:

$$R = a + bv + cv^2$$

where:

R denotes total resistance

a denotes initial constant resistance

b denotes velocity-related constant

c denotes constant that is function of shape, frontal area, and sides of vehicle

v denotes velocity

v^2 denotes velocity, squared

Essentially, resistance differs between vehicle's contact point and surface of roadway. Metal wheels on metal rails have a significant advantage of overcoming resistance compared to rubber-tyred wheels on any road surface (railway – 0.001g at 10 miles per hour (16 km/h) and 0.024g at 60 miles per hour (97 km/h); truck – 0.009g at 10 miles per hour (16 km/h) and 0.090 at 60 miles per hour (97 km/h)). In terms of cargo capacity combining speed and size being moved in a day:

- human – can carry 100 pounds (45 kg) for 20 miles (32 km) per day, or 1 tmi/day (1.5 tkm/day)

- horse and wheelbarrow – can carry 4 tmi/day (5.8 tkm/day)
- horse cart on good pavement – can carry 10 tmi/day (14 tkm/day)
- fully utility truck – can carry 20,000 tmi/day (29,000 tkm/day)
- long-haul train – can carry 500,000 tmi/day (730,000 tkm/day) Most trains take 250–400 trucks off the road, thus making the road more safe.

In terms of the horsepower to weight ratio, a slow-moving barge requires 0.2 horsepower per short ton (0.16 kW/t), a railway and pipeline requires 2.5 horsepower per short ton (2.1 kW/t), and truck requires 10 horsepower per short ton (8.2 kW/t). However, at higher speeds, a railway overcomes the barge and proves most economical.

As an example, a typical modern wagon can hold up to 113 tonnes (125 short tons) of freight on two four-wheel bogies. The track distributes the weight of the train evenly, allowing significantly greater loads per axle and wheel than in road transport, leading to less wear and tear on the permanent way. This can save energy compared with other forms of transport, such as road transport, which depends on the friction between rubber tyres and the road. Trains have a small frontal area in relation to the load they are carrying, which reduces air resistance and thus energy usage.

In addition, the presence of track guiding the wheels allows for very long trains to be pulled by one or a few engines and driven by a single operator, even around curves, which allows for economies of scale in both manpower and energy use; by contrast, in road transport, more than two articulations causes fishtailing and makes the vehicle unsafe.

Energy Efficiency

Considering only the energy spent to move the means of transport, and using the example of the urban area of Lisbon, electric trains seem to be on average 20 times more efficient than automobiles for transportation of passengers, if we consider energy spent per passenger-distance with similar occupation ratios. Considering an automobile with a consumption of around 6 l/100 km (47 mpg_{imp}; 39 mpg_{US}) of fuel, the average car in Europe has an occupancy of around 1.2 passengers per automobile (occupation ratio around 24%) and that one litre of fuel amounts to about 8.8 kWh (32 MJ), equating to an average of 441 Wh (1,590 kJ) per passenger-km. This compares to a modern train with an average occupancy of 20% and a consumption of about 8.5 kW·h/km (31 MJ/km; 13.7 kW·h/mi), equating to 21.5 Wh (77 kJ) per passenger-km, 20 times less than the automobile.

Usage

Due to these benefits, rail transport is a major form of passenger and freight transport in many countries. It is ubiquitous in Europe, with an integrated network covering virtually the whole continent. In India, China, South Korea and Japan, many millions use trains as regular transport. In North America, freight rail transport is widespread and heavily used, but intercity passenger rail transport is relatively scarce outside the Northeast Corridor, due to increased preference of other modes, particularly automobiles and airplanes. South Africa, northern Africa and Argentina have extensive rail networks, but some railways elsewhere in Africa and South America are isolated lines. Australia has a generally sparse network befitting its population density but has some areas with significant networks, especially in the southeast. In addition to the previously existing east-west transcontinental line in Australia, a line from north to south has been constructed. The highest railway in the world is the line to Lhasa, in Tibet, partly running over permafrost territory. Western Europe has the highest railway density in the world and many individual trains there operate through several countries despite technical and organizational differences in each national network.

Social and Economic Benefits

Japanese Shinkansen

Railways are central to the formation of modernity and ideas of progress. Railways contribute to social vibrancy and economic competitiveness by transporting multitudes of customers and workers to city centres and inner suburbs. Hong Kong has recognized rail as "the backbone of the public transit system" and as such developed their franchised bus system and road infrastructure in comprehensive alignment with their rail services. China's large cities such as Beijing, Shanghai, and Guangzhou recognize rail transit lines as the framework and bus lines as the main body to their metropolitan transportation systems. The Japanese Shinkansen was built to meet the growing traffic demand in the "heart of Japan's industry and economy" situated on the Tokyo-Kobe line.

German soldiers in a railway car on the way to the front in August 1914.
The message on the car reads Von München über Metz nach Paris.
(From Munich via Metz to Paris).

During much of the 20th century, rail was an invaluable element of military mobilization, allowing for the quick and efficient transport of large numbers of reservists to their mustering-points, and infantry soldiers to the front lines. However, by the 21st century, rail transport – limited to locations on the same continent, and vulnerable to air attack – had largely been displaced by the adoption of aerial transport.

Railways channel growth towards dense city agglomerations and along their arteries, as opposed to highway expansion, indicative of the U.S. transportation policy, which incents development of suburbs at the periphery, contributing to increased vehicle miles travelled, carbon emissions, development of greenfield spaces, and depletion of natural reserves. These arrangements revalue city spaces, local taxes, housing values, and promotion of mixed use development.

Subsidies

Asia

China

In 2014, total rail spending by China was $130 billion and is likely to remain at a similar rate for the rest of the country's next Five Year Period (2016–2020).

India

The Indian railways are subsidized by around ₹400 billion (US$6.0 billion), of which around 60% goes to commuter rail and short-haul trips. It is the fourth largest railway network in the world comprising 119,630 kilometres (74,330 mi) of total track and 92,081 km (57,216 mi) of running track over a route of 66,687 km (41,437 mi) with 7,216 stations at the end of 2015–16.

Europe

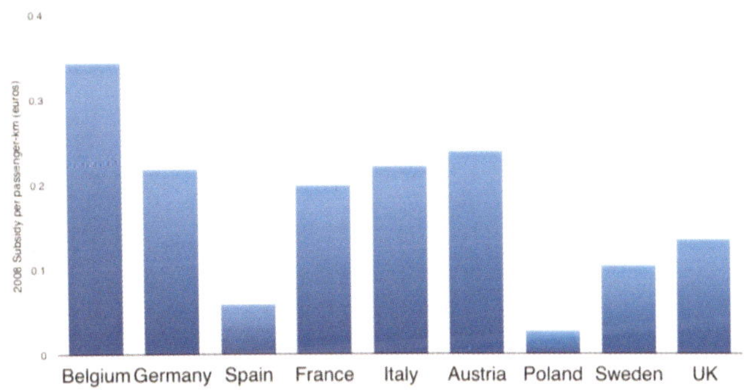

European rail subsidies in euros per passenger-km

Country	Subsidy in billions of Euros	Year
Germany	17.0	2014
France	13.2	2013
Italy	8.1	2009
Switzerland	5.8	2012
Spain	5.1	2015

Country	Subsidy in billions of Euros	Year
United Kingdom	4.5	2015
Belgium	3.4	2008
Netherlands	2.5	2014
Austria	2.3	2009
Denmark	1.7	2008
Sweden	1.6	2009
Poland	1.4	2008
Ireland	0.91	2008

Russia

In total, Russian Roadways receives 90 billion roubles (around US$1,5 billion) annually from the government.

North America

United States

Current subsidies for Amtrak (passenger rail) are around $1.4 billion. The rail freight industry does not receive subsidies.

Advantages of Rail Transport

1. It is relatively safer to travel by train than to travel by road transport or air transport.
2. Rail transport has the ability of conveying heavy and bulky goods; it is also very cheap, safe and comfortable for passengers over a long distance.
3. Rail transport provides a fast and reliable service; trains running on their own tracks can be timed to minimize the chance of congestion and delay.
4. Heavy and bulky goods can be carried as well as large quantities of any cargo.
5. Rail is especially suitable to steady flows of traffic between set places for instance, movement of coal and iron ore between mines and factories.
6. High-speed trains, which carry mainly passengers, help to reduce time spent on travels. Such trains are called Bullet trains in Japan, InterCity 125 in Britain and Express or limited passenger Train in Nigeria.
7. Rails like any other mode of transport helps to open up new and many remote places.
8. Rail is particularly suited to passenger transport since it provides speed, cheapness, comfort and safety; this advantage is greatest for distances under 400km since above distance the completion from air strengthen.

9. Railroad transport offers reasonable good access along its line side although this is dependent upon the number of stations along each route); rail can provide a magnet to industry and stations becomes centres of new urban complexes. For instance, Crewe, Swindon and Peterborough).

10. Rail is a relatively clean form of transport and has only little air pollution released to the atmosphere.

Disadvantages of Rail Transport

1. It involves high construction costs, maintenance costs and operating costs, this remains true whether the system is used or not.
2. It can be expensive to use, especially over short distances or for small amounts of cargo.
3. Rail transport is relatively slower than some other means of transport especially in Africa.
4. Inflexible service since tracks cannot easily be moved and transhipment of cargo is often necessary at termini.
5. Indirect or circuitous routes since lines are determined by relief, hence a slow movement of commodities in areas of rough topography especially across mountain or swampy regions.
6. Traffic flow is variable especially commuter movement around towns and seasonal flows of farm produce; at times, when traffic slack, capital and labour are underemployed.
7. It cannot accommodate awkward loads since cargo must fit both train and routes dimensions (about 3m width).
8. Lines established in the past may become out-dated and a financial burden.
9. Rail gauges differ in many countries. For instances trains going eastwards from Germany or Poland to USSR may have to change their wheels before they cross international border to USSR and even in Africa, rail gauges in all countries are not the same.

Road Transport

Road transport means transportation of goods and personnel from one place to the other on roads. Road is a route between two destinations, which has been either paved or worked on to enable transportation by way of motorised and non-motorised carriages. There are many advantages of road transport in comparison to other means of transport. The investment required in road transport is very less compared to other modes of transport such as railways and air transport. The cost of construction, operating cost and maintaining roads is cheaper than that of the railways.

Road transport compared with other modes of transportation is more flexible, no specialised machinery or technique is necessary for it use so that all countries no matter how backward, possess them in some form. In primitive regions, they may be simple parts suitable only for foot or hoof traffic, but in advance regions, they are more likely to be well made and suitable for heaviest and the most modern vehicles. Although stone-slab surfaces date back to Persian and Roman times, it was not really until the 18th century that roads proper ceased to be mere dirt tracks rutted in summer and water logged in winter. The first metalled surfaces appeared in Britain (designed by such men as Telford and Macadam), but soon spread throughout the civilized world. Vehicles, too improved and carriages became swift, smoother and larger.

The invention of the internal combustion engine gave the greatest incentive of all to all traffic improvements and since then this particular form of transport has changed beyond all recognition. The development of wider, straighter roads has culminated in the appearance of motorways and vehicles can now travel at speeds once unimaginable. Loads have also grown and â€œjuggernautâ€ lorries carrying forty tones are commonplace.

Description

Road transport can be classified as transporting either goods and materials or transporting people. The major advantage of road transport is that it can enable door-to-door delivery of goods and materials and can provide a very cost-effective means of cartage, loading and unloading. Sometimes road transport is the only way for carrying goods and people to and from rural areas which are not catered to by rail, water or air transport. Delivery of goods between cities, towns and small villages is made possible only through road transport. However, in spite of various merits, road transport has some major limitations. For instance, there are more chances of accidents and breakdowns in case of road transport. So, motor transport is not as safe as other means of transport. Road transport is also quite less organised in comparison with other modes. It is irregular and undependable. Rates for road transportation are also unstable and unequal, while the speed in road transport is slow and limited, which is a major drawback. Transporting bulky goods over long distances is also unsuitable and costly. In modern days, road transport has a serious negative impact on the environment. Building roads requires melting of tar or formulation of concrete, which may harm the associated environment. Since roads have been a major enabler of motorised transport, these vehicles also emit a lot of pollution in the form of Nitrogen dioxide, volatile organic compounds, carbon monoxide and various harmful air pollutants, including benzene, which have an adverse respiratory health effects and a serious threat to global warming. While improvisation of roads is a serious topic of research, road transport of the future includes aspects like solar panel roads and cars where solar cells have replaced asphalt or tar, and there are vehicles with electric motors reducing emission. Road transport of the future aims to work on these negativities and turn them around.

Trucking and Haulage

A truck transporting a container on Interstate 95 in South Florida.

B double parked near the Hume Highway

and safety sign on rear

Trucking companies (AE) or haulage companies / hauliers (BE) accept cargo for road transport. Truck drivers operate either independently – working directly for the client – or through freight carriers or shipping agents. Some big companies (e.g. grocery store chains) operate their own internal trucking operations. The market size for general freight trucking was nearly $125 billion in 2010.

In the U.S. many truckers own their truck (rig), and are known as owner-operators. Some road transportation is done on regular routes or for only one consignee per run, while others transport goods from many different loading stations/shippers to various consignees. On some long runs only cargo for one leg of the route (to) is known when the cargo is loaded. Truckers may have to wait at the destination for the return cargo (from).

A bill of lading issued by the shipper provides the basic document for road freight. On cross-border transportation the trucker will present the cargo and documentation provided by the shipper to customs for inspection. This also applies to shipments that are transported out of a free port.

To avoid accidents caused by fatigue, truckers have to keep to strict rules for drivetime and required rest periods. In the United States and Canada, these regulations are known as hours of service, and in the European Union as drivers working hours. One such regulation is the Hours of Work and Rest Periods (Road Transport) Convention, 1979. Tachographs record the times the vehicle is in motion and stopped. Some companies use two drivers per truck to ensure uninterrupted transportation; with one driver resting or sleeping in a bunk in the back of the cab while the other is driving.

Truck drivers often need special licences to drive, known in the U.S. as a commercial driver's license. In the U.K. a large goods vehicle licence is required.

For transport of hazardous materials truckers need a licence, which usually requires them to pass an exam (e.g. in the EU). They have to make sure they affix proper labels for the respective hazard(s) to their vehicle. Liquid goods are transported by road in tank trucks (AE) or tanker lorries (BE) (also road-tankers) or special tankcontainers for intermodal transport. For unpackaged goods and liquids weigh stations confirm weight after loading and before delivery. For transportation of live animals special requirements have to be met in many countries to prevent cruelty to animals. For fresh and frozen goods refrigerator trucks or reefer (container)s are used.

In Australia road trains replace rail transport for goods on routes throughout the center of the country. B-doubles and semi-trailers are used in urban areas because of their smaller size. Low-loader or flat-bed trailers are used to haul containers, containerization, in intermodal transport.

Modern Roads

Highway D1 in Slovakia.

Today, roadways are primarily asphalt or concrete. Both are based on McAdam's concept of stone aggregate in a binder, asphalt cement or Portland cement respectively. Asphalt is known as a flexible pavement, one which slowly will "flow" under the pounding of traffic. Concrete is a rigid pavement, which can take heavier loads but is more expensive and requires more carefully prepared subbase. So, generally, major roads are concrete and local roads are asphalt. Concrete roads are often covered with a thin layer of asphalt to create a wearing surface.

Modern pavements are designed for heavier vehicle loads and faster speeds, requiring thicker slabs and deeper subbase. Subbase is the layer or successive layers of stone, gravel and sand supporting the pavement. It is needed to spread out the slab load bearing on the underlying soil and to conduct away any water getting under the slabs. Water will undermine a pavement over time, so much of pavement and pavement joint design are meant to minimize the amount of water getting and staying under the slabs.

Shoulders are also an integral part of highway design. They are multipurpose; they can provide a margin of side clearance, a refuge for incapacitated vehicles, an emergency lane, and parking space. They also serve a design purpose, and that is to prevent water from percolating into the soil near the main pavement's edge. Shoulder pavement is designed to a lower standard than the pavement in the traveled way and won't hold up as well to traffic, so driving on the shoulder is generally prohibited.

Pavement technology is still evolving, albeit in not easily noticed increments. For instance, chemical additives in the pavement mix make the pavement more weather resistant, grooving and other surface treatments improve resistance to skidding and hydroplaning, and joint seals which were once tar are now made of low maintenance neoprene.

Traffic Control

Disruptions in organized traffic flow can create delays lasting hours.

Nearly all roadways are built with devices meant to control traffic. Most notable to the motorist are those meant to communicate directly with the driver. Broadly, these fall into three categories: signs, signals or pavement markings. They help the driver navigate; they assign the right-of-way at intersections; they indicate laws such as speed limits and parking regulations; they advise of potential hazards; they indicate passing and no passing zones; and otherwise deliver information and to assure traffic is orderly and safe.

Two hundred years ago these devices were signs, nearly all informal. In the late 19th century signals began to appear in the biggest cities at a few highly congested intersections. They were manually operated, and consisted of semaphores, flags or paddles, or in some cases colored electric lights, all modeled on railroad signals. In the 20th century signals were automated, at first with electromechanical devices and later with computers. Signals can be quite sophisticated: with vehicle sensors embedded in the pavement, the signal can control and choreograph the turning movements of heavy traffic in the most complex of intersections. In the 1920s traffic engineers learned how to coordinate signals along a thoroughfare to increase its speeds and volumes. In the 1980s, with computers, similar coordination of whole networks became possible.

In the 1920s pavement markings were introduced. Initially they were used to indicate the road's centerline. Soon after they were coded with information to aid motorists in passing safely. Later, with multi-lane roads they were used to define lanes. Other uses, such as indicating permitted turning movements and pedestrian crossings soon followed.

In the 20th century traffic control devices were standardized. Before then every locality decided on what its devices would look like and where they would be applied. This could be confusing, especially to traffic from outside the locality. In the United States standardization was first taken at the state level, and late in the century at the federal level. Each country has a Manual of Uniform Traffic Control Devices (MUTCD) and there are efforts to blend them into a worldwide standard.

Besides signals, signs, and markings, other forms of traffic control are designed and built into the roadway. For instance, curbs and rumble strips can be used to keep traffic in a given lane and median barriers can prevent left turns and even U-turns.

Toll Roads

Early toll roads were usually built by private companies under a government franchise. They typically paralleled or replaced routes already with some volume of commerce, hoping the improved road would divert enough traffic to make the enterprise profitable. Plank roads were particularly attractive as they greatly reduced rolling resistance and mitigated the problem of getting mired in mud. Another improvement, better grading to lessen the steepness of the worst stretches, allowed draft animals to haul heavier loads.

A *toll road* in the United States is often called a *turnpike*. The term *turnpike* probably originated from the gate, often a simple pike, which blocked passage until the fare was paid at a *toll house* (or *toll booth* in current terminology). When the toll was paid the pike, which was mounted on a swivel, was turned to allow the vehicle to pass. Tolls were usually based on the type of cargo being transported, not the type of vehicle. The practice of selecting routes so as to avoid tolls is called shunpiking. This may be simply to avoid the expense, as a form of economic protest (or boycott), or simply to seek a road less traveled as a bucolic interlude.

Companies were formed to build, improve, and maintain a particular section of roadway, and tolls were collected from users to finance the enterprise. The enterprise was usually named to indicate the locale of its roadway, often including the name of one of both of the termini. The word *turnpike* came into common use in the names of these roadways and companies, and is essentially used interchangeably with *toll road* in current terminology.

In the United States, toll roads began with the Lancaster Turnpike in the 1790s, within Pennsylvania, connecting Philadelphia and Lancaster. In the state of New York, the Great Western Turnpike was started in Albany in 1799 and eventually extended, by several alternate routes, to near what is now Syracuse, New York.

Toll roads peaked in the mid 19th century, and by the turn of the twentieth century most toll roads were taken over by state highway departments. The demise of this early toll road era was due to the rise of canals and railroads, which were more efficient (and thus cheaper) in moving freight over long distances. Roads wouldn't again be competitive with rails and barges until the first half of the 20th century when the internal combustion engine replaces draft animals as the source of motive power.

With the development, mass production, and popular embrace of the automobile, faster and higher capacity roads were needed. In the 1920s limited access highways appeared. Their main characteristics were dual roadways with access points limited to (but not always) grade-separated interchanges. Their dual roadways allowed high volumes of traffic, the need for no or few traffic lights along with relatively gentle grades and curves allowed higher speeds.

The first limited access highways were *Parkways*, so called because of their often park-like landscaping and, in the metropolitan New York City area, they connected the region's system of parks. When the German autobahns built in the 1930s introduced higher design standards and speeds, road planners and road-builders in the United States started developing and building toll roads to similar high standards. The Pennsylvania Turnpike, which largely followed the path of a partially built railroad, was the first, opening in 1940.

After 1940 with the Pennsylvania Turnpike, toll roads saw a resurgence, this time to fund limited access highways. In the late 1940s and early 1950s, after World War II interrupted the evolution of the highway, the US resumed building toll roads. They were to still higher standards and one road, the New York State Thruway, had standards that became the prototype for the U.S. Interstate Highway System. Several other major toll-roads which connected with the Pennsylvania Turnpike were established before the creation of the Interstate Highway System. These were the Indiana Toll Road, Ohio Turnpike, and New Jersey Turnpike.

Interstate Highway System

In the United States, beginning in 1956, Dwight D. Eisenhower National System of Interstate and Defense Highways, commonly called the Interstate Highway System was built. It uses 12 foot (3.65m) lanes, wide medians, a maximum of 4% grade, and full access control, though many sections don't meet these standards due to older construction or constraints. This system created a continental-sized network meant to connect every population center of 50,000 people or more.

By 1956, most limited access highways in the eastern United States were toll roads. In that year, the federal Interstate highway program was established, funding non-toll roads with 90% federal dollars and 10% state match, giving little incentive for states to expand their turnpike system. Funding rules initially restricted collections of tolls on newly funded roadways, bridges, and tunnels. In some situations, expansion or rebuilding of a toll facility using Interstate Highway Program funding resulted in the removal of existing tolls. This occurred in Virginia on Interstate

64 at the Hampton Roads Bridge-Tunnel when a second parallel roadway to the regional 1958 bridge-tunnel was completed in 1976.

Since the completion of the initial portion of the interstate highway system, regulations were changed, and portions of toll facilities have been added to the system. Some states are again looking at toll financing for new roads and maintenance, to supplement limited federal funding. In some areas, new road projects have been completed with public-private partnerships funded by tolls, such as the Pocahontas Parkway (I-895) near Richmond, Virginia.

The newest policy passed by Congress and the Obama Administration regarding highways is the Surface and Air Transportation Programs Extension Act of 2011.

Pneumatic Tires

As the horse-drawn carriage was replaced by the car and lorry or truck, and speeds increased, the need for smoother roads and less vertical displacement became more apparent, and pneumatic tires were developed to decrease the apparent roughness. Wagon and carriage wheels, made of wood, had a tire in the form of an iron strip that kept the wheel from wearing out quickly. Pneumatic tires, which had a larger footprint than iron tires, also were less likely to get bogged down in the mud on unpaved roads.

Advantages of Road Transport

1. Road transport creates flexibility of service since numerous routes and destinations are possible.
2. The maintenance cost involved in road transport is low.
3. It offers directness and door-to-door communication is made possible.
4. It enables great speed and cheapness over short distances, continuous movement is made possible.
5. Road transport carries more passengers and goods within localities because they are normally constructed to connect towns and villages where rail, air and water transports are absent.
6. Wide range of goods can be carried since roads are more able than other systems to handle outsized objects.
7. It offers maximum access along their line sides: maximum scope for intervening opportunity.

Disadvantages of Road Transport

1. Roads and motor vehicles need constant maintenance and this adds greatly to running costs.
2. Road transport links many places more than any other means of transport
3. Most drivers and road users are very careless and as a result of this, a lot of accidents have claim thousands of lives and property.

4. Road transport at times can be very slow due to many factors such as poor maintenance, weather and traffic congestion.
5. Most cities have so many vehicles which ply the road, hence resulting in traffic congestion.
6. Road transport is not suitable for conveying heavy and bulky goods.
7. Motorways are detrimental to the beauty of the countryside and to urban living, motor vehicles cause a lot of air pollution in the city leading to global warming and other environmental hazards.
8. Transport by roads does not operate on schedule and therefore it is very disappointing most times.
9. Goods are not normally weighed and therefore charges are indiscriminately done.

Maritime Transport

Maritime transport relates to the carriage of goods or/and passengers by sea by Maritime transport relates to the carriage of goods or/and passengers by sea by a person for commercial purposes, either in return for payment (i.e. for hire and reward) or on an organization's own account as part of its wider economic activity. Goods transportation refers to the volume of containerized, dry bulk, liquid bulk and roll on-roll off (Ro-Ro) type of cargo handled by the ports while passenger traffic refers to the number of national, international and cruise passenger volumes transported through ports.

Port operations are a necessary tool to enable maritime trade between trading partners. To ensure smooth port operations and to avoid congestion in the harbor it is inevitable to permanently upgrade the port's physical infrastructure, invest in human capital, fostering connectivity of the port and upgrade the port operations to prevailing standards. Hence, port operations can be defined as all policies, reforms and regulations that influence the infrastructure and operations of port facilities including shipping services.

Maritime transport is difficult to negotiate for two reasons. It involves several different kinds of services, from shipping to port facilities. And each of these services is organised in complex ways. Some (including bulk shipping) are relatively competitive, others (such as much of liner shipping) are dominated by rate-setting conferences, bilateral arrangements or monopolies. Some are operated by state enterprises. Some are highly protected with strong domestic political lobbies and long-established labour union practices. Finding appropriate formulas for negotiated liberalisation in these circumstances is not easy.

The negotiations have dealt with the so-called three pillars of the maritime transport sector:

- International shipping: transporting passengers or freight between ports in different countries.
- Auxiliary services such as cargo-handling, storage and warehousing, stevedoring, freight forwarding, customs clearance services, container-station and depot services. Negotiations deal also with foreign operators' rights to establish their own facilities and supply these services.

- Access to and use of port facilities, such as pilotage, towing and tug assistance, provisioning, garbage collection, port captain›s services, and anchorage. Negotiations deal with rights of foreign ships to gain access to these services without discrimination.

A fourth pillar, "multimodal transport" emerged later in the negotiations with a US proposal. Essentially a door-to-door service, this involves the use of one or more "modes" (ie: road, rail, air, or inland water transport) in addition to shipping by sea.

The negotiations have not dealt with domestic coastal shipping known as cabotage (i.e. shipping between ports in the same country), a highly protected sector. Most countries reserve this kind of transportation for national-flag ships. Economists argue that the costs of cabotage protection are extremely high.

Broadly speaking, there are two kinds of shipping: bulk (complete shiploads of commodities and more recently steel, paper and even cars, for example) and liner services (crates, containers, etc of freight, and passengers). Bulk shipping is considered to be more competitive. Liner shipping is traditionally dominated by rate-setting bodies known as conferences. Many countries exempt these conferences from anti-monopoly laws but how this is done varies from country to country.

Seaborne shipping and its international liberalisation is further complicated by the use of flags of convenience such as those of Panama and Liberia. Flag of convenience laws mean ships can be owned by nationals of one country and be registered in another. Most ships in the major ship-owning nations sail under flags of convenience (71% of US ships, 56% of Greek ships, 69% of Japanese ships, 81% of Hong Kong ships, 73% of UK ships and 61% of German ships, according to UNCTAD figures for 1994.) Under GATS, it is the nationality of the owner that counts.

An important additional factor is shipping technology. This is changing so rapidly that in some areas experts say negotiating liberalisation ought to be easier than during the Uruguay Round, even only two years after the round ended. One example of rapid change is in port handling of containerised cargoes. This has allowed movement away from public sector structures towards market-oriented approaches including deregulation, decentralisation, anti-monopoly laws, and privatisation.

Merchant Shipping

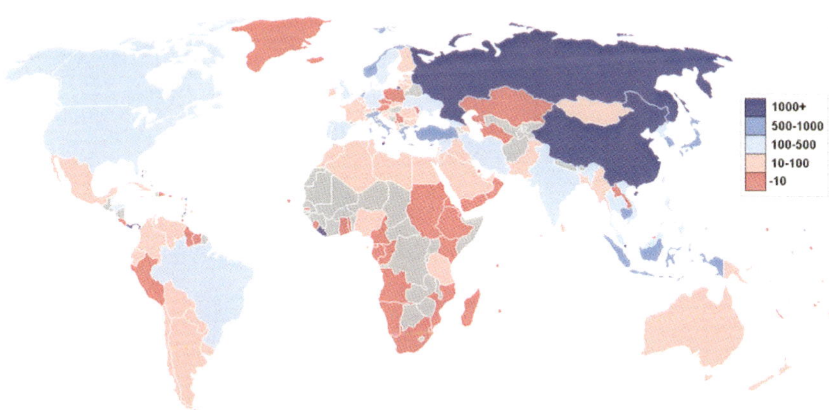

registration of merchant ships (1,000 gross register tons (GRT) and over) per country.

A nation's shipping fleet (merchant navy, merchant marine, merchant fleet) consists of the ships operated by civilian crews to transport passengers or cargo from one place to another. Merchant shipping also includes water transport over the river and canal systems connecting inland destinations, large and small. For example, during the early modern era, cities in the Hanseatic League began taming Northern Europe's rivers and harbors. And, for instance, the Saint Lawrence Seaway connects the port cities on the Great Lakes in Canada and the United States with the Atlantic Ocean shipping routes; while the various Illinois Canals connect the Great Lakes and Canada with New Orleans. Ores, Coal, and grains can travel along the rivers of the American midwest to Pittsburgh, or Birmingham. Professional mariners are merchant seaman, merchant sailor, and merchant mariner, or simply seaman, sailor, or mariners. The terms "seaman" or "sailor" may refer to a member of a country's navy.

According to the 2005 CIA World Factbook, the total number of merchant ships of at least 1,000 gross register tons in the world was 30,936. In 2010, it was 38,988, an increase of 26%. Statistics for individual countries are available at the list of merchant navy capacity by country.

Ships and Watercraft

Ships and other watercraft are used for maritime transport. Types can be distinguished by propulsion, size or cargo type. Recreational or educational craft still use wind power, while some smaller craft use internal combustion engines to drive one or more propellers, or in the case of jet boats, an inboard water jet. In shallow draft areas, such as the Everglades, some craft, such as the hovercraft, are propelled by large pusher-prop fans.

Most modern merchant ships can be placed in one of a few categories, such as:

Bulk carriers, such as the Sabrina I, are cargo ships used to transport bulk cargo items such as ore or food staples (rice, grain, etc.) and similar cargo. They can be recognized by the large box-like hatches on their deck, designed to slide outboard for loading. A bulk carrier could be either dry or wet. Most lakes are too small to accommodate bulk ships, but a large fleet of lake freighters has been plying the Great Lakes and St. Lawrence Seaway of North America for over a century.

Container ships are cargo ships that carry their entire load in truck-size containers, in a technique called containerization. They form a common means of commercial intermodal freight transport. Informally known as "box boats," they carry the majority of the world's dry cargo. Most container ships are propelled by diesel engines, and have crews of between 10 and 30 people. They generally have a large accommodation block at the stern, directly above the engine room.

Tankers are cargo ships for the transport of fluids, such as crude oil, petroleum products, liquefied petroleum gas (LPG), liquefied natural gas (LNG) and chemicals, also vegetable oils, wine and other food - the tanker sector comprises one third of the world tonnage.

Refrigerated ships (usually called Reefers) are cargo ships typically used to transport perishable commodities which require temperature-controlled transportation, mostly fruits, meat, fish, vegetables, dairy products and other foodstuffs

Roll-on/roll-off ships, such as the Chi-Cheemaun, are cargo ships designed to carry wheeled cargo such as automobiles, trailers or railway carriages. RORO (or ro/ro) vessels have built-in ramps which allow the cargo to be efficiently "rolled on" and "rolled off" the vessel when in port. While smaller ferries that operate across rivers and other short distances still often have built-in ramps, the term RORO is generally reserved for larger ocean-going vessels, including pure car/truck carrier (PCTC) ships.

Coastal trading vessels, also known as coasters, ships used for trade between locations on the same island or continent. They are often small and of shallow draft, and sometimes set up as self-dischargers.

Ferries are a form of transport, usually a boat or ship, but also other forms, carrying (or *ferrying*) passengers and sometimes their vehicles. Ferries are also used to transport freight (in lorries and sometimes unpowered freight containers) and even railroad cars. Most ferries operate on regular, frequent, return services. A foot-passenger ferry with many stops, such as in Venice, is sometimes called a waterbus or water taxi. Ferries form a part of the public transport systems of many waterside cities and islands, allowing direct transit between points at a capital cost much lower than bridges or tunnels. Many of the ferries operating in Northern European waters are ro/ro ships.

Cruise ships are passenger ships used for pleasure voyages, where the voyage itself and the ship's amenities are considered an essential part of the experience. Cruising has become a major part of the tourism industry, with millions of passengers each year as of 2006. The industry's rapid growth has seen nine or more newly built ships catering to a North American clientele added every year since 2001, as well as others servicing European clientele. Smaller markets such as the Asia-Pacific region are generally serviced by older tonnage displaced by new ships introduced into the high growth areas. On the Baltic sea this market is served by cruiseferries.

Ocean liner is a passenger ship designed to transport people from one seaport to another along regular long-distance maritime routes according to a schedule. Ocean liners may also carry cargo or mail, and may sometimes be used for other purposes.

Ocean liners are usually strongly built with a high freeboard to withstand rough seas and adverse conditions encountered in the open ocean, having large capacities for fuel, food and other consumables on long voyages. These were the main stay of most passenger transport companies, however, due to the growth of air travel, the passenger ships saw a steady decline. Cruise ships later filled the void and are primarily used by people who still have a love of the sea and offer more amenities compared to the older passenger ships.

Cable layer is a deep-sea vessel designed and used to lay underwater cables for telecommunications, electricity, and such. A large superstructure, and one or more spools that feed off the transom distinguish it.

A tugboat is a boat used to manoeuvre, primarily by towing or pushing other vessels in harbours, over the open sea or through rivers and canals. They are also used to tow barges, disabled ships, or other equipment like towboats.

A dredger (sometimes also called a dredge) is a ship used to excavate in shallow seas or fresh water areas with the purpose of gathering up bottom sediments and disposing of them at a different location.

A barge is a flat-bottomed boat, built mainly for river and canal transport of heavy goods. Most barges are not self-propelled and need to be moved by tugboats towing or towboats pushing them. Barges on canals (towed by draft animals on an adjacent towpath) established the conditions supporting the early industrial revolution in both Europe and the American Northeast but later after they made possible steam locomotive prime movers riding iron rails – after both could grow (and mature) to become commonplace and capable – contended with the railways and were outcompeted in the carriage of people, light freight, and high value items due to the higher speed, falling costs, and route flexibility of rail transport. Carriage of bulk goods also gradually lost ground to freight railways as train capacity and speeds continued to climb. Even underpowered early rail networks could usually reach places only an outrageously expensive canal might be built, and once Iron T-rails and higher powered locomotives became possible, the far cheaper to build railways were unfettered and independent upon water sources, whilst mostly unplagued by the seasonal problems (restricted by icing) of temperate latitude canals which suffered ice and freshet flooding damages with dreary regularity. When floods did affect railways, restoration of services was usually comparatively rapid.

A Multi-purpose ship (sometimes called a general cargo ship) is used to transport a variety of goods from bulk commodities to break bulk and heavy cargoes. To provide maximum trading flexibility they are usually geared and modern examples are fitted for the carriage of containers and grains. Generally they will have large open holds and tweendecks to facilitate the carriage of different cargoes on the same voyage. The crew will be highly competent in the securing of break bulk cargoes and the ship will be equipped with various lashings and other equipment for sea fastening.

Liners and Tramps

A ship may also be categorized as to how it is operated.

A liner will have a regular run and operate to a schedule. The scheduled operation requires that such ships are better equipped to deal with causes of potential delay such as bad weather. They are generally higher powered than tramp ships with better seakeeping qualities, thus they are significantly more expensive to build. Liners are typically built for passenger and container operation though past common uses also included mail and general cargo.

A tramp has no fixed run but will go wherever a suitable cargo takes it. Thus a ship and crew may be chartered from the ship owner to fetch a cargo of grain from Canada to Latvia, the ship may then be required to carry a cargo of coal from Britain to Melanesia. Bulk carriers and some cruise ships are examples of ships built to operate in this manner.

Typical in-transit Times

A cargo ship sailing from a European port to a US one will typically take 10–12 days depending on water currents and other factors. In order to make container ship transport more economical in the face of declining demand for intercontinental shipping, ship operators sometimes reduce cruising speed, thereby increasing transit time, to reduce fuel consumption, a strategy referred to as "slow steaming".

Carbon Emissions

Maritime transport accounts for roughly 3% of global carbon dioxide emissions globally. Although the industry was not a focus of attention of the Paris Climate Accord signed in 2016, the United Nations and the International Maritime Organisation (IMO) have discussed CO_2 emissions goals and limits. In 2018, the industry discussed in London placing limits to cut levels from a benchmark of 2008 carbon dioxide emissions by 50% by the year 2050. Some methods of reducing emissions of the industry include lowering speeds of shipping (which can be potentially problematic for perishable goods) as well as changes to fuel standards.

Maritime Transport Infrastructure

Container port facilities in Newark Bay, seen from Bayonne, New Jersey, United States

For a port to efficiently send and receive cargo, it requires infrastructure: docks, bollards, pilings, cranes, bulk cargo handling equipment, and so on – equipment and organization supporting the role of the facilities. From pier to pier these may differ, one dock handling intermodal transport needs (container-ships linked to rail by cranes); another bulk handling capabilities (such as conveyors, elevators, tanks, pumps) for loading and unloading bulk cargoes like grain, coal, or fuels. Others may be outfitted as passenger terminals or for mixed mode operations.

Generally, Harbors, seaports and marinas all host watercraft, and consist of components such as piers, wharfs, docks and roadsteads.

Air Transport

Air transport is the most recent mode of transport. It is the gift of the 20th century to the world. The two world wars gave a great impetus to the development of air transport in almost all the countries of the world. The peculiar characteristic of air transport is that is does not need a specific surface track for its operations.

It has no physical barriers as in the case of other mode of transport. Political boundaries are also immaterial although it has to observe the requirements of the International Law. The supreme advantage of air transport lies in its quickness.

It is the fastest mode of transport. But the cost of its operation is very high and thus it is suitable for only rich passengers, mails and light and costly cargo. However, in advanced countries like U.S.A., Germany, etc. it offers a tough competition to the railways.

Types of Air Transport

1. Commercial airplanes
2. Helicopters
3. Private planes
4. Blimps
5. Gliders
6. hang gliders
7. Zeppelin
8. parachute

Commercial airplanes: these are the common ways in which people travel through the air, the commercial planes provide a fast means of transportation compared to other modes of transport such as road transport, rail transport and water transport. Airplanes are capable of carrying hundreds of people from on location to another at a time; the seating is sometimes divided into two or four classes. For instance, most domestic flights usually have two classes which are: First Class and Economy Class. While international flights may have up to four classes such as First Class, Club Class, Business Class, Premium Economy and Economy Class.

Helicopters: helicopters are another fast means of air transport; these move people through the air. Helicopters when compared with commercial airplanes are much more limited when it comes to passengerâ€™s space and can only transport a few people at a time, whereas some commercial airplanes can transport hundreds of people at a time.

Private planes: private planes are made to provide transportation service for a single person or at most five (5) people at a time. Private planes range from the smallest Cessna to luxury jets such as the Citation CJ1, which carries up to five people in leather-seated comfort. These planes provide comfort and privacy during the trip and these could travel from one location to another for a business purpose or for pleasure.

Blimps: blimps and hot air balloons are used to transport people for recreational purposes. They cover a limited area that enables tourists to see a location on a larger scope than if they tried to view the area from the ground. Blimps used to be a form of commercial transport but are no longer used for that.

Glider: a glider, which is also called sailplane, Â is a type of Â glider aircraft Â used in the sport of Â gliding or for recreational activity. Sailplanes are aerodynamically streamlined and are capable of gaining altitude when flown in rising air. Â Some modern gliders are made of an Â aluminium, alloyÂ or Â compositeÂ frame covered with synthetic Â sailcloth which forms the wings.

Rocket: a rocket is any vehicle that uses a rocket engine; it includes a missile, spacecraft, aircraft or other vehicle. Rockets have been used at least since the 13th century for small-scale military applications and recreational displays. Rockets work more in space than in the atmosphere, the engines work by action and reaction of pushing the rockets forward simply by expelling their exhaust in the opposite direction at high speed, and can therefore work in the vacuum of space.

Zeppelin: a Zeppelin was a type of rigid airship named after the German Count Ferdinand von Zeppelin, it consists of a cigar-shaped, trussed, and covered frame supported by internal gas cells. Count Ferdinand von Zeppelin designed Zeppelin in the early 20th century. Zeppelins almost look like blimps but they differ by two points: Zeppelins have a metal skeleton with a rigid covering, and they use hydrogen gas to float. These two elements made zeppelins larger when compared to blimps.

Parachute: this is a cloth canopy that is filled with air and allows a person, package or a heavy object attached to it to descend slowly when dropped from an aircraft, or which is released from the rear of an aircraft on landing to act as a brake. Back then, the military developed parachuting technology as a way of saving aircrews from emergencies aboard balloons and aircraft in flight, and later as a way of delivering soldiers to the battlefield. Nowadays, parachuting is performed as a recreational activity and as a competitive sport in various places.

Development of Air Transport

Air transport is the newest means of transport; this means of transport was introduced in 1903 but developed into full means of transporting people and freights in the 1930s. The greatest of air transportation started after the Second World War (WW11). This means of transportation can be used for both domestic and international flights.

Of all transport development of the 20th century, those in air transport have been the most striking ones. Who would have thought that when the Wright brothers made their historic flight in 1903, that aircraft would become one of the most important means of passengers transport within just three generations? Improvements have indeed been rapid: jet engines replaced propellers, radar was introduced, the size of aircraft has grown to â€œjumboâ€ proportions; supersonic speeds have been achieved and vertical take-off is now possible.

Today, across the world, air transport is used extensively both for passengers and freight. Broadly, there are two types of services: Those operating for particular purposes on and â€˜ad hocâ€™ basis and those operating on regular schedules. Into the first category, you will have the flights. (For example, those for tourist in summer and for mineral deposits between inaccessible mines and industrial regions), however, into the latter category would come those services run by British airways and the other world airlines. Increasingly, the routes of both types radiate from the developed regions and especially from the great capital cities and industrial centres of the world. Whenever possible they mark the shortest distance between places and thus use the â€˜Great Circleâ€™ routes to economize both time and fuel. Many cities including London, New york, Cairo and Bangkok possess major international airports and many others are developing their own international airport facilities. In low areas, where there is difficult terrain, air transport provides the only communication possible and assumes a correspondingly greater significance.

Domestic and International Flights

Air travel can be separated into two general classifications: national/domestic and international flights. Flights from one point to another within the same country are called domestic flights. Flights from a point in one country to a point within a different country are known as international flights. Travelers can use domestic or international flights in either private or public travel.

Air Travel

Travel class on an airplane is usually split into a two, three or four class model service. US Domestic flights usually have two classes: Economy Class and a Domestic First Class partitioned into cabins. International flights may have up to four classes: Economy Class; Premium Economy; Business Class or Club Class; and First Class.

Most air travel starts and ends at a commercial airport. The typical procedure is check-in; border control; airport security baggage and passenger check before entering the gate; boarding; flying; and pick-up of luggage and - limited to international flights - another border control at the host country's border.

On commercial airlines for long journey or some time on short journeys the air travel consists of two flights. It could have many reasons i.e there is break at the route or the flight has passengers of different destinations. Usually the two way flight to a route is cheaper than the one way route.

Environmental Effects

Modern aircraft consume less fuel per person and mile travelled than cars when fully booked. This argument in favor of air travel is counterweighted by two facts:

1. The distances travelled are often significantly larger and will not replace car travel but instead add to it, and
2. Not every flight is booked out.

Instead, the scheduled flights are predominant, resulting in a far worse fuel efficiency. According to the ATAG, flights produced 781 million tonnes (769 million long tons) of the greenhouse gas CO_2 in 2015 globally, as compared to an estimated total of 36 billion tonnes (35 billion long tons) anthropogenic CO_2.

Health Effects

Deep vein thrombosis (DVT) is the third-most common vascular disease, next to stroke and heart attack. It is estimated that DVT affects one in 5,000 travellers on long flights. Risk increases with exposure to more flights within a short time frame and with increasing duration of flights.

During flight, the aircraft cabin pressure is usually maintained at the equivalent of 6,000–8,000 ft (1,800–2,400 m) above sea level. Most healthy travelers will not notice any effects. However, for travelers with cardiopulmonary diseases (especially those who normally require supplemental oxygen), cerebrovascular disease, anemia, or sickle cell disease, conditions in an aircraft can exacerbate underlying medical conditions. Aircraft cabin air is typically dry, usually 10%–20% humidity, which can cause dryness of the mucous membranes of the eyes and airways.

Advantages of Air Transport

1. Air transport is a fast and efficient system which is especially suited to passenger traffic, offering comfort and high quality service; today long distance passenger movements are made almost entirely due to air transport.
2. Air transport involves the use of direct routes.
3. It is suited to high quality, expensive and perishable cargoes for which speed is essential; aircraft may also carry freight too small to interest the providers of other transport systems.
4. It can often reach areas inaccessible to other modes of transport. For instance, areas of in hospitable environment such as Central Brazil and Northern Canada.
5. It offers complete freedom of movement and this provides potentially flexible routes and services (although most air movements are channelled along strict routes and political factors may restrict both landing and flight paths).

Disadvantages of Air Transport

1. Air transport is very expensive and as a result of this, it disqualifies all heavy, bulky or low-value cargoes.

2. Aircraft are expensive to build and operate; it requires elaborate provisions in the form of airport facilities, controlling systems and maintenance.
3. It depends on weather conditions and therefore can provide and unreliable services.
4. As aircraft becomes larger, the number of airports capable of being used is declining. This reduces flexibility. There is a growing difficulty of finding suitable sites for airport building as well.
5. Some airports are far away from urban centres and this offsets the advantage of speed and convenience.
6. It offers no access between termini and therefore minimum potential for intervening opportunity.
7. Airports use up valuable land and aircraft cause a lot of noise and air pollution.

Intermodal Passenger Transport

Both passenger and cargo transport are believed to benefit from Intermodal transport.

Intermodal transport is a complex transport system. Even definitions of Intermodal transport vary and sometimes contradict. Intermodal transport is not only given a range of different names, but also a wide variety of descriptions are in use, making the debate / discussion on the subject difficult and confusing1 . To avoid entering in the details of this discussion, this book therefore adopts the Intermodal transport definition of Prof. Dr. Gerhard Muller that reads: "the concept of transporting passengers and freight on two or more different transport modes in such a way that all parts of the transportation process, including the exchange of information, are efficiently connected and coordinated"

It is a kind of transportation modal involving the multiple mode of transportation like air, water, land via Aircraft, ship, rails and trucks. This system can be used for the transportation of freight and passengers both.

While in a transportation system for freight the cargo is packed in containers and that container is transported to destination via multiple modes like ships, aircrafts, rails, trucks etc. without removing the cargo from the container. Since there is not much handling of container required this system is very safe and secure and requires less handling time.

This leads to very economic mode of transportation.

Intermodal transportation for passenger works in similar way. The passenger uses different mode to reach to their destination. We use this mode in our daily life. For example a person reaching to an airport by using the bus, train or taxi service is going through the intermodal passenger transportation system.

Urban mixed-mode Commuting

Public transportation systems such as train or metro systems have the most efficient means and highest capacity to transport people around cities. Therefore, mixed-mode commuting in the urban environment is largely dedicated to first getting people onto the train network and once off the train network to their final destination.

Automobile to Public Transport Nodes

Although automobiles are conventionally used as a single-mode form of transit, they also find use in a variety of mixed-mode scenarios. They can provide a short commute to train stations, airports, and piers, where all-day "park and ride" lots are often available. Used in this context, cars offer commuters the relative comfort of single-mode travel, while significantly reducing the financial and environmental costs.

Taxicabs and Rental cars also play a major role in providing door-to-door service between airports or train stations and other points of travel throughout urban, suburban, and rural communities.

(Automobiles can also be used as the centerpiece of a multi-mode commute, with drivers resorting to walking or cycling to their final destination. Commuters to major cities take this route when driving is convenient, but parking options at the destination are not readily available.)

Park and Ride

Kiss and Ride in Vienna

Transport planners often try to encourage automobile commuters to make much of their journey by public transport. One way of doing this is to provide car parking places at train or bus stations where commuters can drive to the station, park their cars and then continue on with their journey on the train or bus: this is often called Park-and-Ride.

Similar to Park-and-Ride is what is often termed "Kiss-and-Ride". Rather than drive to the train or bus station and park the commuter is driven to the station by a friend or relative (parent, spouse etc.) The "Kiss" refers to the peck on the cheek as the commuter exits the car. *Kiss-and-Ride* is usually conducted when the train/bus/ferry station is close to home, so that the driver dropping the commuter off has a short journey to and from home.

Bus to Public Transport Nodes

Many large cities link their railway network to their bus network. This enables commuters to get to places that are not serviced directly by rail as they are often considered to be too far for walking.

Chicago's Jefferson Park Transit Center is an intermodal hub for bus and train commuters.

Feeder buses are a specific example of this; feeder buses service local neighbourhoods by taking travellers from their homes to nearby train stations which is important if the distances are too far to comfortably walk; at the end of the working day the buses take the travellers home again. Feeder buses work best when they are scheduled to arrive at the railway station shortly before the train arrives allowing enough time for commuters to comfortably walk to their train, and on the commuters' return journey buses are scheduled to arrive shortly after the train arrives so that the buses are waiting to take the commuters home. If train and bus services are very frequent then this scheduling is unimportant as the commuter will in any case have a very short wait to interchange.

Bike-and-Ride

A *Parkiteer* bicycle parking station at Sunshine railway station, Melbourne in Australia

All around the world bicycles are used to get to and from train and other public transportation stations; this form of intermodal passenger transport is often called "Bike-and-Ride" (n.b. the "ride" here confusingly refers to the part of the trip on public transport; the term is constructed along the lines of *Park-and-Ride*). To safeguard against theft or vandalism of parked bicycles at these train, bus, and ferry stations, "bike and ride" transport benefits greatly from secure bicycle parking facilities such as bicycle parking stations being available.

Some train, bus, and ferry systems allow commuters to take their bicycles aboard, allowing cyclists to ride at both ends of the commute, though sometimes this is restricted to off-peak travel periods:

in such cases, folding bicycles may be permitted where regular bicycles are not. In some cities, bicycles are permitted aboard trains and buses.

Public bicycles at Euston station

In some cities a public bicycle rental programme allows commuters to take a public bike between the public transport station and a docking station near their origin or destination.

The use of "*Bike-and-Ride*" instead of a car can cut costs for fuel and parking, and some families no longer need to own and operate multiple cars. Environmental benefits can also increase (i.e. less pollution) and reduced traffic congestion can deliver significant cost savings to the city and local government.

Many transit agencies have begun installing bike racks on the front of buses, as well as in the interior of buses, trains, and even on ferries. These transit bike racks allow cyclists the ability to ride their bicycle to the bus/train/ferry, take the mode of transportation, then ride again to their final destination. These types of racks combined with increased bike infrastructure and bike parking have made bike commuting a frequent topic of discussion by cities and local government.

Inter-regional Mixed Mode Commuting

Intermodal Passenger Transport Involving Air Travel

Airport rail link

Many cities have extended subway or rail service to major urban airports. This provides travellers with an inexpensive, frequent and reliable way to get to their flights as opposed to driving or being driven, and contending with full up parking, or taking taxis and getting caught in traffic jams on the way to the airport. Many airports now have some mass transit link, including London, Sydney, Munich, Hong Kong, Vancouver, Philadelphia, and New York City (JFK).

Airport–ferry connection

At the Hong Kong International Airport, ferry services to various piers in the Pearl River Delta are provided. Passengers from Guangdong can use these piers to take a flight at the Airport, without passing through customs and immigration control, effectively like having a transit from one flight to another. The Airport is well-connected with expressways and an Airport Express train service. A seaport and logistics facilities will be added in the near future. Kansai International Airport is also connected to Kobe Airport with ferries. The Toronto Island ferry connects Billy Bishop Toronto City

Airport to mainland Toronto, where passengers can connect to the Toronto streetcar system or with airport shuttle buses which transports to bus, subway and rail connections at Union Station.

Automobiles on Trains

Several passenger rail systems offer services that allow travelers to bring their automobiles with them. These usually consist of automobile carrying wagons attached to normal passenger trains, but some special trains operate solely to transport automobiles. This is particularly of use in areas where trains may travel but automobiles cannot, such as the Channel Tunnel. Another system called NIMPR is designed to transport electric vehicles on high speed trains.

Trains on Boats

A train ferry is a ship designed to carry railway vehicles. While usually used to carry freight vehicles, passenger cars can also be carried. In other places passengers move between passenger cars to a passenger ferry.

Train–ferry Connection

Prior to the widespread use of automobiles, the San Francisco Bay Area featured a complex network of ferry services which connected numerous interurban and streetcar systems in the North and East Bay to the San Francisco Ferry Building, where several city streetcar lines began service. The opening of the rail-carrying San Francisco–Oakland Bay Bridge and automotive Golden Gate Bridge almost entirely supplanted these services.

Sonoma–Marin Area Rail Transit commuter rail is expected to feature a connection with the Golden Gate Ferry and service to San Francisco Ferry Building at Larkspur Landing. The Hercules Station & Terminal is to be the first direct Amtrak-to-ferry transit hub in the San Francisco Bay.

The Staten Island Railway, while operated by the Metropolitan Transportation Authority, does not have a physical connection to the rest of New York City's rail network. As such, transfers to Manhattan are facilitated by the free Staten Island Ferry.

Transfer Facilities

In recent years, an increasing emphasis has been placed on designing facilities that make such transfers easier and more seamless. These are intended to help passengers move from one mode (or form) of transportation to another. An intermodal station may service air, rail, and highway transportation for example.

In some cases, facilities were merged or transferred into a new facility, as at the William F. Walsh Regional Transportation Center in Syracuse, New York or South Station in Boston, Massachusetts. In other cases new facilities, such as the Alewife Station In Cambridge, Massachusetts were built from the start to emphasize intermodalism.

Regional transit systems in the United States often include regional intermodal transit centers that incorporate multiple types of rail and bus services alongside Park and Ride amenities. Until the completion of San Francisco Transbay Transit Center, the Millbrae Intermodal Terminal in

California is the largest intermodal transit center west of the Mississippi which includes direct, on-platform connections between BART, the Bay Area's regional rail system, Caltrain the San Francisco Peninsula's commuter rail, and SamTrans, the regional bus service for San Mateo County. The uniqueness of this transfer facility is that turnstiles are located on the platforms between rail services in addition to on a separate concourse to allow for direct transfers. Millbrae Intermodal Terminal is also planned to be incorporated into the California High-Speed Rail project as one of two stations between San Francisco and San Jose.

Assessment

Advantages

Mixed mode commuting combines the benefits of walking, bicycle commuting, or driving with the benefits of rapid transit while offsetting some of the major disadvantages of each. The use of a bicycle can, for example, make an (inexpensive compared to a car) 20 mile light-rail or suburban rail journey attractive even if the endpoints of the journey each sit 1 mile out from the stations: the 30 minutes walking time becomes 8 minutes bicycling.

As in the example above, location plays a large role in mixed mode commuting. Rapid transit such as express bus or light rail may cover most of the distance, but sit too far out from commute endpoints. At 3 mph walking, 2 miles represents about 40 minutes of commute time; whereas a bicycle may pace 12 mph leisurely, cutting this time to 10 minutes. When the commuter finds the distance between the originating endpoint (e.g. the home) and the destination (e.g. the place of employment) too far to be enjoyable or practical, commute by car or motorcycle to the station may remain practical, as long as the commute from the far end station to the destination is practical by walking, a carry-on cycle, or another rapid transit such as a local or shuttle bus.

In general, locations close to major transit such as rail stations carry higher land value and thus higher costs to rent or purchase. A commuter may select a location further out than practical walking distance but not more than practical cycling distance to reduce housing costs. Similarly, a commuter can close an even further distance quickly with an ebike, motorcycle, or car, allowing for the selection of a more preferred living area somewhat further from the station than would be viable by walking or simple bicycle.

Other cost advantages of mixed mode commuting include lower vehicle insurance via Pay As You Drive programs; lower fuel and maintenance costs; and increased automobile life. In the most extreme cases, a mixed-mode commuter may opt to car share and pay only a small portion of purchase, fuel, maintenance, and insurance, or to live car-free. These cost benefits are offset by costs of transit, which can vary. A Maryland MTA month pass valid for MTA Light Rail, Metro Subway, and City Bus costs $64, while a month pass for the Baltimore to DC MARC costs $175.00 and a DC MetroRail 7 day pass costs $47 totaling $182.

Disadvantages

The effectiveness of a mixed-mode commute can be measured in many ways: speed to destination, convenience, security, environmental impact, and proximity to mass transit are all factors. Because mixed-mode commutes rely on a certain degree of coordination, scheduling issues with mass transit can often be an issue. For example, a sometimes-late train can be an annoyance, and an often-late train can make a commute impractical.

Weather can also be a factor. Even when the use of an automobile is involved, the transition from one mode of transportation to another often exposes commuters to the elements. As a result, multi-mode commuters often travel prepared for inclement weather.

Transportation Demand Management

Transportation demand management (TDM) has become a hot topic in recent years. This is thanks in large part to the many ways in which technology has made using and promoting smart transportation choices easier. From a conceptual standpoint, TDM is concerned with the ways in which people make optimal use of locally available transportation resources, with a strong focus on getting people out of single-occupancy vehicles and into more efficient modes of commuting.

In essence, TDM balances the people-focused and infrastructure-focused ways in which problems like traffic congestion, infrastructure costs, parking challenges, and environmental impact can be managed or reduced. Strategically, it functions on two major levels:

At the surface ...

At the surface level, TDM aims to provide information, incentives, resources, and support to people who want to make the best possible use of available transportation options. These alternatives include public transit, carpooling, vanpooling, ridesharing, walking, and cycling. Some conceptual models also include telecommuting as a TDM topic.

Going deeper ...

On a deeper level, TDM is also concerned with urban design and municipal planning. Specifically, TDM strategies can be used to encourage broader engagement with transportation alternatives, and guide local residents to use them more often. At this level, key concepts include walkability indices and "complete streets," sustainability, urban livability, and the integrated management of key transportation corridors.

While practically all major cities use TDM principles to inform policy decisions and development projects to some degree, it is still a relatively underused tool despite its powerful reach and transformative potential.

Background

The term TDM has its origins in the United States in the 1970s and 1980s, and is linked to the economic impacts of the sharp increase in oil prices during the 1973 oil crisis and the 1979 energy crisis. When long lines appeared at gas stations, it became self-evident that alternatives to single-occupancy commuter travel needed to be provided in order to save energy, improve air quality, and reduce peak period congestion.

Traffic congestion on Interstate 5, at Los Angeles, California.

The concepts of TDM borrowed from mainstream transport planning in Europe, which had never been based on assumptions that the private car was the best or only solution for urban mobility. For example, the Dutch Transport Structure Scheme has since the 1970s required that demand for additional vehicle capacity be met only "if the contribution to societal welfare is positive" and since 1990 has included an explicit target to halve the rate of growth in vehicle traffic.

Some cities outside Europe have also consistently taken a demand management approach to transport and land use planning, notably Curitiba, Brazil, Portland, Oregon, USA, Arlington, Virginia, USA, and Vancouver, Canada.

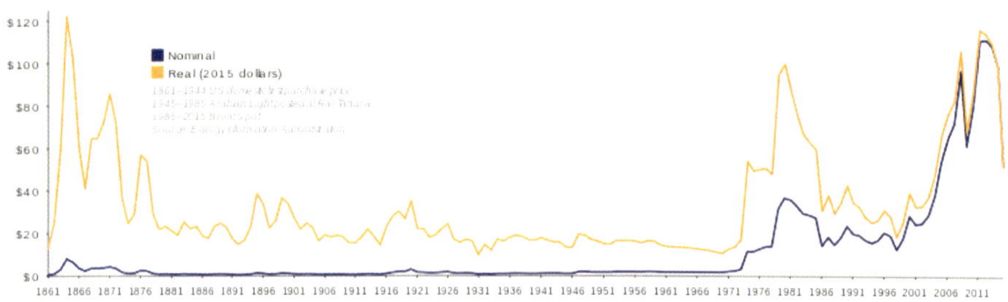

Oil price trend, 1861–2007, both nominal and adjusted to inflation.

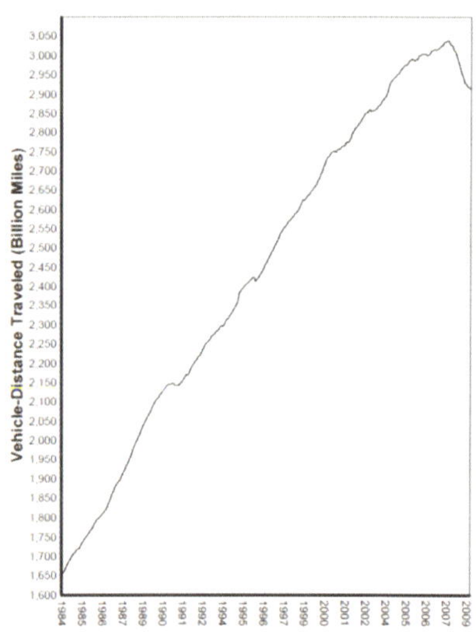

Vehicle miles travelled in the United States

Relatively low and stable oil prices during the 1980s and 1990s led to significant increases in vehicle travel, both directly because people chose to travel by car more often and for greater distances, and indirectly because cities developed tracts of suburban housing, distant from shops and from workplaces, now referred to as urban sprawl. Trends in freight logistics, including a movement from rail and coastal shipping to road freight and a requirement for just in time deliveries, meant that freight traffic grew faster than general vehicle traffic.

Because vehicle travel was increasing rapidly from 1980–2000, it follows that (with a few exceptions) the techniques of demand management were not widely or successfully applied during this period. Small-scale projects to provide alternatives to single occupant commuter travel were common, but generally were led from outside the mainstream of transport planning. However many of the techniques in the demand management toolbox were developed during this period.

The British Government's White Paper on Transport marked a change in direction. In the introduction to the White Paper, Prime Minister Tony Blair stated that

We recognise that we cannot simply build our way out of the problems we face. It would be environmentally irresponsible - and would not work.

A companion document to the White Paper called "Smarter Choices" researched the potential to scale up the small and scattered sustainable transport initiatives then occurring across Britain, and concluded that the comprehensive application of these techniques could reduce peak period car travel in urban areas by over 20%.

A similar study by the United States Federal Highway Administration, was also released in 2004 and also concluded that a more proactive approach to transportation demand was an important component of overall national transport strategy.

Terminology

Since 2010 transportation professionals have suggested that the TDM is widely misunderstood simply as a collection of vaguely related initiatives, and that this misunderstanding is constraining the true potential of the concept. TDM practitioners have found that TDM is far more effective when framed as a philosophical approach which over time becomes a cornerstone of sustainable urban transport systems. A new paradigm in transport planning, internationally recognised as TDM, appears to be emerging which embraces concepts such as "mobility management" and "active travel management" under its umbrella.

Crucial to the delivery of a sustainable urban transport system is integrating the TDM philosophy into urban transport planning, as well as the daily management and operation of transport services and infrastructure. It appears that managing travel demand has largely been compartmentalized as a set of "soft measures" to promote sustainable travel options or programs to promote and offer shared ride arrangements. Demand management means different things to different disciplines. For example: to Information Technology (IT) specialists, managing demand is new technology to provide information; to operations managers, managing demand is controlling the flow onto highways; to economists, it is pricing the system to find equilibrium with capacity; to marketers, it is promoting innovative campaigns; and to many policymakers TDM remains a largely unknown entity.

The concept has become confused as each discipline has tried to mold the concept to their set of tools. This "silo" thinking inhibits the kind of policy integration that is needed to develop a sustainable urban transport solution strategy. There remains much confusion as to what a sustainable transport system would comprise. It is helpful therefore to consider different approaches to sustainable transport along a spectrum of viewpoints, ranging from weak to strong sustainability. Generally efforts to address the impact of transport on climate change to date have been largely focused on technology. The impact of this technological-led approach has been very limited in the transport sector. TDM has the potential to move the transport sector from a position of weak to strong sustainability by combining behavior-change with technology improvements.

In this context transport demand management is understood as a much broader concept. Implicit in the use of the term is the assumption that it is accompanied by the implementation of sustainable mobility, introduction of full cost pricing and organizational or structural measures to ensure a broad range of complementary interventions work effectively together to realize the benefits of sustainable transport. It is the unifying philosophy of TDM, not specific measures associated with it, that underpin the policy objective of a more sustainable system of transport. This philosophy of managing demand accepts that meeting unfettered demand for travel is impractical and that therefore the system needs to be managed. That demand for travel needs to be managed by:

- Expanding the supply and availability of (more sustainable) alternatives;
- Controlling demand for the use unsustainable modes;
- Providing incentives and rewards for undertaking sustainable travel habits; and
- Imposing full-cost pricing on the use of the automobile.

Arguments in Favor

The need to manage travel demand has now become urgent for a number of converging reasons.

Oil prices have now passed the previous peak in 1980, and 95% of all energy used in transport is oil. Vehicle travel in the United States, which has been rising steadily since records began, started to level out before the fuel price increases and is now in decline. Part of this decline is likely to be people making fewer trips, with potentially far-reaching economic and social consequences. Countries and cities where the car is one of many travel choices are more likely to prosper, as people can choose to drive less but are still able to travel by transit, cycle safely, walk to local shops and facilities, or choose to work or study from home.

Transport systems are responsible for 23% of energy-related greenhouse gas emissions, and are increasing at a faster rate than any other energy using sector. Demand management is central to the effort to reduce greenhouse gas emissions from urban transportation,

Increases in vehicle travel are linked to a range of health problems including poor urban air quality, road injuries and fatalities, and reduced physical activity. The World Health Organisation stated in 2003:

We are concerned that current patterns of transport, which are dominated by motorised road transport, have substantial adverse impacts on health.

The efficacy of expanded roadways in managing traffic congestion is increasingly under challenge. Much of the traffic on new or expanded roads has been shown to be induced.

A growing sustainable transport movement is mobilising public demand for investment in safer, more livable cities with a greater range of travel choices.

Demand Management Toolbox

There is a broad range of TDM measures, including:

- Transportation management associations: leverage public and private funds to increase the use of ridesharing and other commuting options that reduce traffic congestion and improve air quality
- Including or improving pedestrian-oriented design elements, such as short pedestrian crossings, wide sidewalks and street trees.
- Requiring users of parking to pay the costs directly, as opposed to sharing the costs indirectly with others through increased rents and tax subsidies.
- Including and improving public transportation infrastructure, such as subway entrances, bus stops and routes.
- Subsidizing transit costs for employees or residents.
- Bicycle-friendly facilities and environments, including secure bike storage areas and showers.
- Providing active transportation (AT) facilities including bike lanes and multi-use trails.
- Providing traveler information tools, including intelligent transportation system improvements, mobile and social applications, wayfinding tools, and other methods for promoting non-SOV modes
- Flex-time work schedules with employers to reduce congestion at peak times
- Active traffic management
- Congestion pricing during peak hours.
- Road space rationing or alternate-day travel by restricting travel based on license plate number, at certain times and places.
- Workplace travel plans
- Roadspace reallocation, aiming to re-balance provision between private cars which often predominate due to high spatial allocations for roadside parking, and for sustainable modes.
- Time, distance and place (TDP) road pricing, where road users are charged based on when, where and how much they drive. Some transportation experts believe TDP pricing is an integral part of the next generation in transportation demand management.

Objectives of Transportation Demand Management Strategies

Fundamentally, Transportation Demand Management is about more than just managing the way people get around; it's about the overall health and wellness of communities. As such, the field has evolved to include a number of specific objectives, all of which are supported by the use of better methods of transportation. Examples of major TDM objectives include:

Reducing traffic congestion

Getting cars off the road is one of the most common and most immediate goals of transportation demand management. When fewer vehicles are competing for road space, traffic moves more quickly, average commute times come down, and the environmental impact of idling is reduced.

Conserving energy and reducing emissions

The damage caused by vehicle emissions and greenhouse gases is undeniable. It's a major contributor to problems like climate change and environmental degradation. Therefore, getting people to make better use of shared transportation options is one of the most important ways in which communities can do their part to encourage greener thinking.

Improving community health and fitness levels

TDM can also lead to better levels of health and fitness among community members by encouraging people to be more active as they move around town. Improving the walkability of cities and adding cycling infrastructure such as dedicated bike lanes are two of the most important ways TDM strategies can be used to promote healthier and more active lifestyles.

Achieving equity

One of the major problems associated with prioritizing the needs of drivers is that it leads to inequality. As more resources are dedicated to the infrastructure needs of single-occupancy vehicles, those who don't drive or are unable to afford their own car suffer. The best approaches to TDM seek to create a level playing field, in which the interests of all community members are taken into account in equal measure.

Boosting urban livability

Studies have shown that community-oriented modes of transportation can lead to significant improvements in personal satisfaction and happiness. People are more engaged when they are active stakeholders in the communities they live in. By improving social quality for residents, commuters, and visitors alike, TDM helps improve the overall livability of cities.

Solving parking problems

High demand for parking can cause major headaches in urban areas, leading to major increases in traffic congestion at peak times. TDM offers streamlined approaches to the creation and management of parking infrastructure by making it more accessible and affordable, thus reducing spillover rates and improving traffic flow.

Enhancing community safety

The more cars that are on the road, the more pedestrian, cyclist, and motorist safety becomes an issue. One of the key ancillary benefits of reducing urban traffic congestion is improvements in community safety. As such, it's a central concern of any comprehensive TDM strategy.

Helping commuters based in rural areas

People who work in cities but live in areas with low population densities tend to commute by car. Most often this is simply because alternatives aren't readily available. Examples of programs that encourage rural-based commuters to make better use of alternative options include free or inexpensive parking at mass transit stations located outside central business districts, carpool lots along common routes for easier ride sharing, and increases to comfort and security features at these locations.

Making alternative transportation more affordable

One of the most effective ways to get people to leave their cars at home is to make TDM-preferred modes of transportation more affordable than solo driving. Making communities less dependent on single-occupancy vehicles also has a trickle-down effect. It also reduces the amount of resources people have to earmark for their transportation needs. This, in turn, supports other important TDM objectives, including better livability, improved equity and community safety, and reductions in traffic congestion and environmental impact.

Various methods can be used to achieve these objectives. TDM-friendly policy decisions and legislation are essential, but elements like incentives for using sustainable transportation, disincentives for driving, education and information accessibility are also important.

Enacting Positive Change through Transportation Demand Management

The most effective and successful approaches to transportation demand management draw on the combined power of a range of strategies to achieve the aforementioned objectives. Common strategies include:

- Making information about locally available alternatives to single-occupancy vehicles easy to access and use.
- Marketing the benefits of commuting alternatives to business owners so they encourage their employees to use them.
- Supporting a diverse range of transportation options, including not only public transit, but also carpooling, vanpooling, ridesharing, walking, cycling, and others.
- Incentivizing the use of smart, sustainable commuting methods.
- Discouraging the use of single-occupancy vehicles by introducing tolls and surcharges that increase the cost gap between solo driving and other options.
- Introducing limits on driver accessibility to key commercial areas of cities with high levels of traffic congestion.

- Capping single-occupancy vehicle trips, or increasing the average passenger density of privately operated vehicles.

Changing the culture and attitude of commuters naturally comes with difficult challenges. It isn't something that can be achieved overnight. Yet, as TDM becomes an increasingly visible element of urban living, awareness naturally rises and people become more mindful about the impact of their transportation decisions. One of the most powerful ways to continue to foster and nurture change is to leverage technology. Modern user-facing TDM tools make finding and using better methods of transportation easier.

Sustainable Transport

Sustainable Transport is sometimes known as Green Transport and it is any form of transport that does not use or rely on dwindling natural resources. Instead it relies on renewable or regenerated energy rather than fossil fuels that have a finite life expectancy. For this reason it is said to have a low or a negative effect on the environment since it makes use of energy sources that are sustainable.

Walking, cycling and sailing are excellent examples of sustainable transport.

Until relatively recent times very little consideration was given to the sustainable sources of energy for transport, or for the efficient use of clean fuels. In many parts of the developing world this is still the case and high levels of pollution continue to be tolerated.

It is estimated that every year various forms of transport consume between 20% and 25% of the world's energy and this significantly contributes to the increasingly high levels of greenhouse gases that continue to be released into the atmosphere.

It can't Last Forever

There are now around 750 million cars in the world and predictions are that if the present trend continues, the number will double in the next 30 years. All but a very few are powered by petroleum-based fuel.

In 1900 there were just a few thousand cars in the world but as numbers increased, so did the demand for oil. Estimates suggest that during the 20thcentury about half of the world's accessible oil deposits were used up.

It is only relatively recently that people began to be concerned about future oil supplies. A combination of political unrest in major oil producing areas like the Middle East, combined with the fact that reserves are running out in areas such as the US, has heightened this concern and had the combined effect of focusing the mind on seeking alternative energy sources, and pushing up prices.

Added to that, during the last few years people have simply become more environmentally conscious. Whereas before, in many people's eyes the resources of the world were simply there to be made use of, there is now an almost universal appreciation of the fact that the world's resources will not last for ever and that once they are gone they are gone for good.

The term sustainable transport came into use as a logical follow-on from sustainable development, and is used to describe modes of transport, and systems of transport planning, which are consistent with wider concerns of sustainability. There are many definitions of the sustainable transport, and of the related terms sustainable transportation and sustainable mobility. One such definition, from the European Union Council of Ministers of Transport, defines a sustainable transportation system as one that:

- Allows the basic access and development needs of individuals, companies and society to be met safely and in a manner consistent with human and ecosystem health, and promotes equity within and between successive generations.
- Is affordable, operates fairly and efficiently, offers a choice of transport mode, and supports a competitive economy, as well as balanced regional development.
- Limits emissions and waste within the planet's ability to absorb them, uses renewable resources at or below their rates of generation, and uses non-renewable resources at or below the rates of development of renewable substitutes, while minimizing the impact on the use of land and the generation of noise.

Sustainability extends beyond just the operating efficiency and emissions. A life-cycle assessment involves production, use and post-use considerations. A cradle-to-cradle design is more important than a focus on a single factor such as energy efficiency.[

Environmental Impact

The Bus Rapid Transit of Metz uses a diesel-electric hybrid driving system, developed by Belgian Van Hool manufacturer.

Transport systems are major emitters of greenhouse gases, responsible for 23% of world energy-related GHG emissions in 2004, with about three quarters coming from road vehicles. Currently 95% of transport energy comes from petroleum. Energy is consumed in the manufacture as well as the use of vehicles, and is embodied in transport infrastructure including roads, bridges and railways.

The first historical attempts of evaluating the Life Cycle environmental impact of vehicle is due to Theodore Von Karman. After decades in which all the analysis has been focused on emending the Von Karman model, Dewulf and Van Langenhove have introduced an model based on the second law of thermodynamics and exergy analysis. Chester and Orwath, have developed a similar model based on the first law that accounts the necessary costs for the infrastructure.

The environmental impacts of transport can be reduced by reducing the weight of vehicles, sustainable styles of driving, reducing the friction of tires, encouraging electric and hybrid vehicles, improving the walking and cycling environment in cities, and by enhancing the role of public transport, especially electric rail.

Green vehicles are intended to have less environmental impact than equivalent standard vehicles, although when the environmental impact of a vehicle is assessed over the whole of its life cycle this may not be the case. Electric vehicle technology has the potential to reduce transport CO_2 emissions, depending on the embodied energy of the vehicle and the source of the electricity. The primary sources of electricity currently used in most countries (coal, gas, oil) mean that until world electricity production changes substantially, private electric cars will result in the same or higher production of CO2 than petrol equivalent vehicles. The Online Electric Vehicle (OLEV), developed by the Korea Advanced Institute of Science and Technology (KAIST), is an electric vehicle that can be charged while stationary or driving, thus removing the need to stop at a charging station. The City of Gumi in South Korea runs a 24 km roundtrip along which the bus will receive 100 kW (136 horsepower) electricity at an 85% maximum power transmission efficiency rate while maintaining a 17 cm air gap between the underbody of the vehicle and the road surface. At that power, only a few sections of the road need embedded cables. Hybrid vehicles, which use an internal combustion engine combined with an electric engine to achieve better fuel efficiency than a regular combustion engine, are already common. Natural gas is also used as a transport fuel. Biofuels are a less common, and less promising, technology; Brazil met 17% of its transport fuel needs from bioethanol in 2007, but the OECD has warned that the success of biofuels in Brazil is due to specific local circumstances; internationally, biofuels are forecast to have little or no impact on greenhouse emissions, at significantly higher cost than energy efficiency measures.

In practice there is a sliding scale of green transport depending on the sustainability of the option. Green vehicles are more fuel-efficient, but only in comparison with standard vehicles, and they still contribute to traffic congestion and road crashes. Well-patronised public transport networks based on traditional diesel buses use less fuel per passenger than private vehicles, and are generally safer and use less road space than private vehicles. Green public transport vehicles including electric trains, trams and electric buses combine the advantages of green vehicles with those of sustainable transport choices. Other transport choices with very low environmental impact are cycling and other human-powered vehicles, and animal powered transport. The most common green transport choice, with the least environmental impact is walking.

Transport on rails boasts an excellent efficiency.

Transport and Social Sustainability

Cities with overbuilt roadways have experienced unintended consequences, linked to radical drops in public transport, walking, and cycling. In many cases, streets became void of "life." Stores, schools, government centers and libraries moved away from central cities, and residents who did not flee to the suburbs experienced a much reduced quality of public space and of public services. As schools were closed their mega-school replacements in outlying areas generated additional traffic; the number of cars on US roads between 7:15 and 8:15 a.m. increases 30% during the school year.

A tram in Melbourne, Australia

Yet another impact was an increase in sedentary lifestyles, causing and complicating a national epidemic of obesity, and accompanying dramatically increased health care costs.

Cities

Futurama, an exhibit at the 1939 New York World's Fair, was sponsored by General Motors and showed a vision of the City of Tomorrow.

Cities are shaped by their transport systems. In The City in History, Lewis Mumford documented how the location and layout of cities was shaped around a walkable center, often located near a port or waterway, and with suburbs accessible by animal transport or, later, by rail or tram lines.

In 1939, the New York World's Fair included a model of an imagined city, built around a car-based transport system. In this "greater and better world of tomorrow", residential, commercial and industrial areas were separated, and skyscrapers loomed over a network of urban motorways. These ideas captured the popular imagination, and are credited with influencing city planning from the 1940s to the 1970s.

Interstate 10 and Interstate 45 near downtown Houston, Texas

The popularity of the car in the post-war era led to major changes in the structure and function of cities. There was some opposition to these changes at the time. The writings of Jane Jacobs, in particular The Death and Life of Great American Cities provide a poignant reminder of what was lost in this transformation, and a record of community efforts to resist these changes. Lewis Mumford asked "is the city for cars or for people?" Donald Appleyard documented the consequences for communities of increasing car traffic in "The View from the Road" (1964) and in the UK, Mayer Hillman first published research into the impacts of traffic on child independent mobility in 1971. Despite these notes of caution, trends in car ownership, car use and fuel consumption continued steeply upward throughout the post-war period.

Mainstream transport planning in Europe has, by contrast, never been based on assumptions that the private car was the best or only solution for urban mobility. For example, the Dutch Transport Structure Scheme has since the 1970s required that demand for additional vehicle capacity only be met "if the contribution to societal welfare is positive", and since 1990 has included an explicit target to halve the rate of growth in vehicle traffic. Some cities outside Europe have also consistently linked transport to sustainability and to land-use planning, notably Curitiba, Brazil, Portland, Oregon and Vancouver, Canada.

There are major differences in transport energy consumption between cities; an average U.S. urban dweller uses 24 times more energy annually for private transport than a Chinese urban resident, and almost four times as much as a European urban dweller. These differences cannot be explained by wealth alone but are closely linked to the rates of walking, cycling, and public transport use and to enduring features of the city including urban density and urban design.

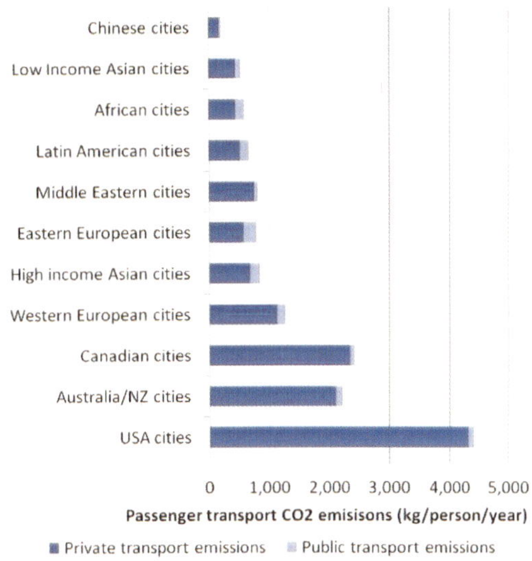

Greenhouse gas emissions from transport vary widely, even for cities of comparable wealth. Source: UITP, Mobility in Cities Database

A bypass the Old Town in Szczecin, Poland

The cities and nations that have invested most heavily in car-based transport systems are now the least environmentally sustainable, as measured by per capita fossil fuel use. The social and economic sustainability of car-based transportation engineering has also been questioned. Within the United States, residents of sprawling cities make more frequent and longer car trips, while residents of traditional urban neighbourhoods make a similar number of trips, but travel shorter distances and walk, cycle and use transit more often. It has been calculated that New York residents save $19 billion each year simply by owning fewer cars and driving less than the average American. A less car intensive means of urban transport is carsharing, which is becoming popular in North America and Europe, and according to The Economist, carsharing can reduce car ownership at an estimated rate of one rental car replacing 15 owned vehicles. Car sharing has also begun in the developing world, where traffic and urban density is often worse than in developed countries. Companies like Zoom in India, eHi in China, and Carrot in Mexico, are bringing car-sharing to developing countries in an effort to reduce car-related pollution, ameliorate traffic, and expand the number of people who have access to cars.

The European Commission adopted the Action Plan on urban mobility on 2009-09-30 for sustainable urban mobility. The European Commission will conduct a review of the implementation of the Action Plan in the year 2012, and will assess the need for further action. In 2007, 72% of the European population lived in urban areas, which are key to growth and employment. Cities need efficient transport systems to support their economy and the welfare of their inhabitants. Around 85% of the EU's GDP is generated in cities. Urban areas face today the challenge of making transport sustainable in environmental (CO_2, air pollution, noise) and competitiveness (congestion) terms while at the same time addressing social concerns. These range from the need to respond to health problems and demographic trends, fostering economic and social cohesion to taking into account the needs of persons with reduced mobility, families and children.

Policies and Governance

Sustainable transport policies have their greatest impact at the city level. Outside Western Europe, cities which have consistently included sustainability as a key consideration in transport and land use planning include Curitiba, Brazil; Bogota, Colombia; Portland, Oregon; and Vancouver, Canada. The state of Victoria, Australia passed legislation in 2010 – the Transport Integration Act – to compel its transport agencies to actively consider sustainability issues including climate change impacts in transport policy, planning and operations.

Many other cities throughout the world have recognised the need to link sustainability and transport policies, for example by joining Cities for Climate Protection.

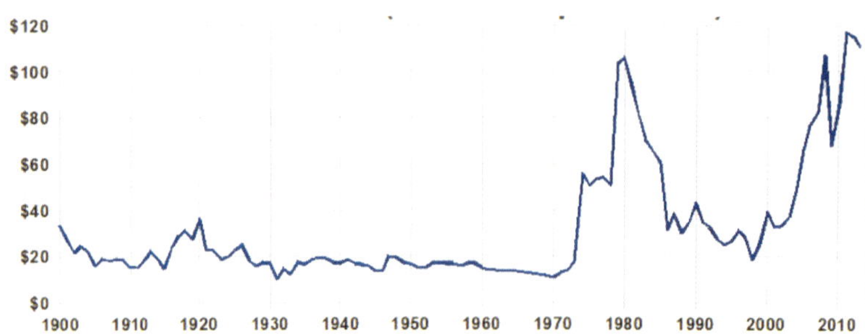

Oil price trend, both nominal and adjusted to inflation

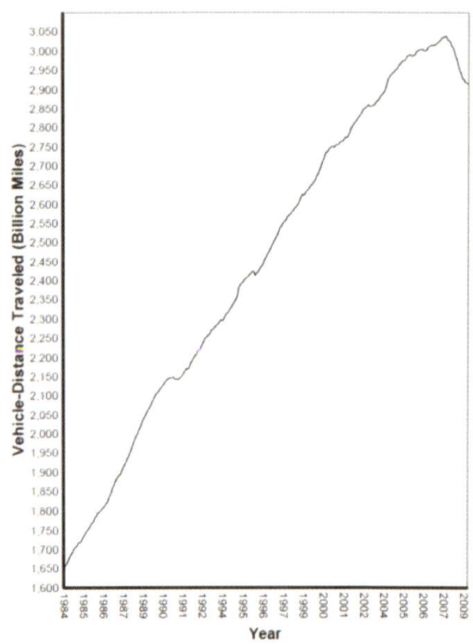

Vehicle-miles traveled in the United States

Community and Grassroots Action

Sustainable transport is fundamentally a grassroots movement, albeit one which is now recognised as of citywide, national and international significance.

Whereas it started as a movement driven by environmental concerns, over these last years there has been increased emphasis on social equity and fairness issues, and in particular the need to ensure proper access and services for lower income groups and people with mobility limitations, including the fast-growing population of older citizens. Many of the people exposed to the most vehicle noise, pollution and safety risk have been those who do not own, or cannot drive cars, and those for whom the cost of car ownership causes a severe financial burden.

An organization called Greenxc started in 2011 created a national awareness campaign in the United States encouraging people to carpool by ride-sharing cross country stopping over at various destinations along the way and documenting their travel through video footage, posts and photography. Ride-sharing reduces individual's carbon footprint by allowing several people to use one car instead of everyone using individual cars.

Recent Trends

Car travel increased steadily throughout the twentieth century, but trends since 2000 have been more complex. Oil price rises from 2003 have been linked to a decline in per capita fuel use for private vehicle travel in the USA, Britain and Australia. In 2008, global oil consumption fell by 0.8% overall, with significant declines in consumption in North America, Western Europe, and parts of Asia. Other factors affecting a decline in driving, at least in America, include the retirement of Baby Boomers who now drive less, preference for other travel modes (such as transit) by younger age cohorts, the Great Recession, and the rising use of technology (internet, mobile devices) which have made travel less necessary and possibly less attractive.

Greenwashing

The term *green transport* is often used as a greenwash marketing technique for products which are not proven to make a positive contribution to environmental sustainability. Such claims can be legally challenged. For instance Norway's consumer ombudsman has targeted automakers who claim that their cars are "green", "clean" or "environmentally friendly". Manufacturers risk fines if they fail to drop the words. The Australian Competition and Consumer Commission (ACCC) describes *green* claims on products as *very vague, inviting consumers to give a wide range of meanings to the claim, which risks misleading them.* In 2008 the ACCC forced a car retailer to stop its *green* marketing of Saab cars, which was found by the Australian Federal Court as *misleading*.

Objectives

The EU Directorate-General for Transport and Energy (DG-TREN) has launched a programme which focusses mostly on Urban Transport. Its main measures are:

<table>
<tr>
<td>• Clean fuels and vehicles
○ Hybrid Vehicles
○ Sailing ships
○ Biodiesel
○ Biogas/CNG
○ Electric Vehicles
○ LPG
○ Hydrogen vehicle
○ Human-powered transport
○ Animal-powered transport
• Sustainable (green) transport infrastructure
○ greenways and foreshoreways
○ Bikeways
○ Busways
○ Railways
• Access restrictions
○ Access management / Enforcement</td>
<td>• Collective passenger transport
○ Public transport
○ Bus services
○ Rail transport
○ Intermodal transfers
○ Integrated ticketing
○ Marketing
○ Park & Ride
○ Demand responsive transport
○ Accessible transport systems
○ Paratransit
○ Bus rapid transit
○ Quality of service
○ Security, including Transit police
• Travel information
○ Public transport route planners (intermodal journey planner)</td>
<td>• Soft measures
○ Travel plan
○ Walking school bus
○ Travel blending
○ Personalised travel plan
• Transport management
○ Transportation demand management
○ Transit oriented development
○ Walkability
○ New urbanism and New pedestrianism
○ TDM Toolbox
• Sustainable Freight Transport
○ Clean vehicles / clean fleet
○ Intermodal freight transport
○ Dry port
○ Fleet management</td>
</tr>
<tr>
<td>○ Car Restricted Zones / Living Streets
○ Multifunctional areas
○ Parking Management
○ Pedestrian zone
○ Traffic calming / Speed reduction
• Integrated pricing strategies
○ Congestion pricing
○ Integrated ticketing
○ Parking Management</td>
<td>• Less car intensive lifestyle
○ Car pooling
○ Car sharing
○ Car/ driver licence exit strategies
○ Cycling
○ Bike sharing
○ More info on bike sharing systems</td>
<td>○ Route planning
○ Transportation management system</td>
</tr>
</table>

Alternatives

In many areas there have been moves to encourage people to use public transport networks rather than travelling by car. Cycling has also gained in popularity, both as a means of maintaining fitness, but also because it is a cheap and often swift form of transport.

There have been moves to introduce cleaner and more fuel-efficient means of urban transport. There have been experiments using different types of fuel and there has been much effort and progress made in the development of so-called ''green vehicles''.

Hybrid vehicles have been developed as an alternative to conventional cars. These are gaining in popularity. Generally they are of a hybrid design, combining an internal combustion engine with an electric engine. Other vehicles run on natural gas and some are electrically powered.

Hybrid electric vehicle (HEV).

Biofuel: Biofuels have also become very popular in some areas. Bioethanol is made by fermenting plant materials and biodiesel is made from vegetable oils, animal fats or recycled grease. In 2008 biofuels provided 1.8% of the world's transport fuel. One of the problems is that bioethanol production relies on the cultivation of large amounts of plant material often on land that was formally tropical rain forest, or on land that would otherwise have been used to produce food.

Solar power is another possible future option, but this will depend on a technical leap in the conversion rates of the photovoltaic (PV) cells that convert sunlight into electricity. Battery power would be required at nighttime or when there was no sun during daylight periods.

More detail on solar powered cars and solar transportation as a method of sustainable transport.

Hydrogen can be used to power future transportation. This might in fact be the power of the future because hydrogen is the most common element in the Universe. Power can either be through the use of electric motors powered by fuel cell technology or by improved internal combustion engines. In both cases emissions would be zero. Hydrogen power is currently prohibitively expensive, but progress is being made in the technology to achieve this. A big challenge is to source the hydrogen from renewable resources. Honda has produced the first 'commercial' hydrogen powered vehicle in the form of the Honda FCX Clarity, although this has limited availability.

Nuclear power was once advocated as the clean alternative to coal, oil and gas for generating electricity and for producing hydrogen through electrolysis. A number of accidents at nuclear plants, together with the inherent problem of dealing with large amounts of radioactive waste, has led many countries to reconsider this method of power generation.

Looking to the Future

Transport is a vital part of our everyday life. The prevailing 20thcentury view was to rely on fossil fuels for most of our transport needs and to ignore any consequences. In the 21stcentury we have woken up to the fact that there is a need for us to care for the world around us.

References

- Wolf, D E; Schreckenberg, M; Bachem, A (June 1996). "Traffic and Granular Flow". Traffic and Granular Flow. WORLD SCIENTIFIC: 1. doi:10.1142/9789814531276. ISBN 9789810226350
- Burney RE, Hubert D, Passini L, Maio R (1995). "Variation in air medical outcomes by crew composition: a two-year follow-up". Ann Emerg Med. 25 (2): 187–92. doi:10.1016/s0196-0644(95)70322-5. PMID 7832345
- "Road Infrastructure Strategic Framework for South Africa". National Department of Transport (South Africa). Archived from the original on 27 September 2007. Retrieved 24 March 2007
- Lewandowski, Krzysztof (2015). "New coefficients of rail transport usage" (PDF). International Journal of Engineering and Innovative Technology (IJEIT). 5 (6): 89–91. ISSN 2277-3754
- S., Kerner, Boris (2004). The Physics of Traffic : Empirical Freeway Pattern Features, Engineering Applications, and Theory. Berlin, Heidelberg: Springer Berlin Heidelberg. ISBN 9783540409861. OCLC 840291446
- "Major Roads of the United States". United States Department of the Interior. 2006-03-13. Archived from the original on 13 April 2007. Retrieved 24 March 2007
- "The Inception of the English Railway Station". Architectural History. SAHGB Publications Limited. 4: 63–76. 1961. doi:10.2307/1568245. JSTOR 1568245
- Cook, R. M.: "Archaic Greek Trade: Three Conjectures 1. The Diolkos", The Journal of Hellenic Studies, Vol. 99 (1979), pp. 152–155 (152)
- Nelson, Donna C., Editor (2000). "Intelligent Transportation Primer". Institute of Transportation Engineers, Washington, D.C. pp. 10-1. ISBN 0-935403-45-0
- van den Hoorn, T and B van Luipen (2003). "National and Regional Transport Policy in the Netherlands" (PDF). Archived from the original (PDF) on 2011-07-24. Retrieved 2008-07-27
- Heather L. MacLean and Lester B. Lave University of Toronto(2003). "OECD's Economic Assessment of Biofuel Support Policies". Environmental Science & Technology. 37 (23): 5445. doi:10.1021/es034574q. Retrieved 1 March 2018

2

Transportation Engineering: Planning and Management

Transportation engineering refers to the application of technology and engineering to the operation, planning and management of the facilities or infrastructure required for any mode of transportation. The aim of this chapter is to explore the fundamental concepts central to this field, such as urban and public transport planning, traffic analysis zone, access management, bicycle transportation planning and engineering, etc.

Transportation Engineering

Transportation Engineering is the application of scientific principles and technologies to the planning, design, analysis, operation, and management of transportation systems for a full range of modes of transportation, e.g. highway, urban transit, air, rail and water, in order to provide for the movement of people and goods from one place to another in a safe and efficient manner. As a sub-discipline of Civil Engineering, this specialty is organized around society's need to provide an adequate transportation infrastructure and involves broad interaction with other disciplines.

Transportation planning requires to understand and model travel behavior in order to forecast trips, assess the necessity of new infrastructures, and propose investment programs and transportation policies to mitigate future congestion and other issues. An example of transportation planning includes studying the necessity of a new highway in a specific region based on population growth, plans for new settlements, and forecasted congestion.

After planning for new infrastructures, transportation designs are required to guide the implementation of specific facilities. They involve the determination of important attributes associated with transportation infrastructures (size, location, geometry, and capacity, among others). An example of transportation design for a new highway includes definition of its specific location, number of lanes, dimensions, curvature, traffic capacity, vertical alignment, pavement thickness, among other features. Transportation designs are key to guide the construction of a transportation facilities.

Construction and maintenance combines knowledge from other areas in civil engineering in order to implement and preserve the designed facilities. An example of transportation construction includes the cost estimation, definition of supplies, and scheduling of activities required to build a specific highway design. An example of maintenance, which is related to the concept of asset management, includes the development of repaving plans based on pre-defined performance metric (traffic intensity, environmental conditions, and infrastructure deterioration, among others).

Transportation operations and management involve the development of traffic engineering solutions (subarea of transportation engineering) to guarantee the smooth movement of traveling entities on the transportation network. Examples of transportation operations and management include the development of signs, signals, marking, movements, tolls, intelligent transportation systems, and other solutions that facilitate the movement of vehicles on a roadway system.

Transportation engineers aim for solutions that are safe, efficient, comfortable, economical, sustainable, and environmentally friendly. Transportation is an important and interdisciplinary area of civil engineering composed by multiple subareas including, but not limited to: traffic, human factors, geometric design, pavements, transit, freight, unimodal and intermodal transportation, policies, intelligent transportation systems, modeling and simulation, among others.

Highway Engineering

Engineers in this specialization:

- Handle the planning, design, construction, and operation of highways, roads, and other vehicular facilities as well as their related bicycle and pedestrian realms.
- Estimate the transportation needs of the public and then secure the funding for the project.
- Analyze locations of high traffic volumes and high collisions for safety and capacity.
- Use civil engineering principles to improve the transportation system.
- Utilizes the three design controls which are the drivers, the vehicles, and the roadways themselves.

Railroad Engineering

Railway engineers handle the design, construction, and operation of railroads and mass transit systems that use a fixed guideway (such as light rail or even monorails). Typical tasks would include determining horizontal and vertical alignment design, station location and design, and construction cost estimating. Railroad engineers can also move into the specialized field of train dispatching which focuses on train movement control.

Railway engineers also work to build a cleaner and safer transportation network by reinvesting and revitalizing the rail system to meet future demands. In the United States, railway engineers work with elected officials in Washington, D.C. on rail transportation issues to make sure that the rail system meets the country's transportation needs.

Port and Harbor Engineering

Port and harbor engineers handle the design, construction, and operation of ports, harbors, canals, and other maritime facilities.

Airport Engineering

Airport engineers design and construct airports. Airport engineers must account for the impacts and demands of aircraft in their design of airport facilities. These engineers must use the analysis of predominant wind direction to determine runway orientation, determine the size of runway border and safety areas, different wing tip to wing tip clearances for all gates and must designate the clear zones in the entire port.

Transportation Planning

Transportation planning plays a fundamental role in the state, region or community's vision for its future. It includes a comprehensive consideration of possible strategies; an evaluation process that encompasses diverse viewpoints; the collaborative participation of relevant transportation-related agencies and organizations; and open, timely, and meaningful public involvement.

Transport planning is defined as planning required in the operation, provision and management of facilities and services for the modes of transport to achieve safer, faster, comfortable, convenient, economical and environment-friendly movement of people and goods. It is a prediction of usage demand in future travel and to ensure all the necessary facilities and services to cater to that demand. Transport planning is highly essential in shaping cities, enabling economic activities, promoting community interaction, and enhancing quality of life. It is also essential for sustainable development and ensuring safe accessibility at various levels for all individuals.

Description: Under transportation planning, different practices such as urban development, urban transport policies, legislative activities, funding bodies and project management come together. It involves various agencies and organizations for successful transportation planning. Governing bodies actually examine the demographic characteristics and travel patterns for a given area, and check how these characteristics will change over a given period of time. Alternative improvements are then evaluated for the transportation system. Moreover, it requires a lot of coordination between transport system and land use authorities. The aim for all these bodies is todevice a plan which incorporates efficient energy use and lesser pollution of various means. The activity involves alternative modes of transport that will enhance its efficiency. Transportation planning must cover all aspects of city life such as economic development, quality of life, health of public and environment and thereby supporting long-term ecological balance. For this transportation planners and engineers always focus on the efficient movement of people and goods across the country.

Models and Sustainability

Chicago Transit Authority Chicago 'L' trains use elevated tracks for a portion of the system, known as the Loop, which is in the Chicago Loop community area. It is an example of the siting of transportation facilities that results from transportation planning.

A bypass the Old Town in Szczecin, Poland

Transportation planning, or transport planning, has historically followed the rational planning model of defining goals and objectives, identifying problems, generating alternatives, evaluating alternatives, and developing plans. Other models for planning include rational actor, transit oriented development, satisficing, incremental planning, organizational process, collaborative planning, and political bargaining.

Planners are increasingly expected to adopt a multidisciplinary approach, especially due to the rising importance of environmentalism. For example, the use of behavioural psychology to persuade drivers to abandon their automobiles and use public transport instead. The role of the transport planner is shifting from technical analysis to promoting sustainability through integrated transport policies. For example, in Hanoi, the increasing number of motorcycles is responsible for not only environmental damage but also slowing down economic growth. In the long run, the plan is to reduce traffic through a change in urban planning. Through economic incentives and attractive alternatives experts hope to lighten traffic in the short run.

Accessibility (Metric)

In transportation planning, accessibility refers to the ease of reaching destinations. Academics have disputed how the term "ease" should be defined and measured. People who are in places that are highly accessible can reach many other activities or destinations quickly, people in inaccessible places can reach fewer places in the same amount of time.

A measure that is often used is to measure accessibility in a traffic analysis zone i is:

$$Accessibility_i = \sum_j Opportunities_j \times f\left(C_{ij}\right) \text{ where:}$$

i = index of origin zones

j = index of destination zones

$f\left(C_{ij}\right)$ = function of generalized travel cost (so that nearer or less expensive places are weighted more than farther or more expensive places).

For a non-motorized mode of transport, such as walking or cycling, the generalized travel cost may include additional factors such as safety or gradient.

Transport for London utilize a calculated approach known as Public Transport Accessibility Level (PTAL) that uses the distance from any point to the nearest public transport stops, and service frequency at those stops, to assess the accessibility of a site to public transport services.

There are three ways in spatial planning to improve accessibility and they are respectively mobility, connectivity and proximity. They are considered the means to the end which is accessibility.

In the United Kingdom, transport planning has traditionally been a branch of civil engineering. In the 1950s and the 1960s, it was generally believed that the motor car was an important element in the future of transport as economic growth spurred on car ownership figures. The role of the transport planner was to match motorway and rural road capacity against the demands of economic growth. Urban areas would need to be redesigned for the motor vehicle or impose traffic containment and demand management to mitigate congestion and environmental impacts. The policies were popularised in a 1963 government publication, *Traffic in Towns*. The contemporary Smeed Report on congestion pricing was initially promoted to manage demand but was deemed politically unacceptable. In more recent times, the approach has been caricatured as "predict and provide" to predict future transport demand and provide the network for it, usually by building more roads.

The publication of Planning Policy Guidance 13 in 1994 (revised in 2001), followed by A New Deal for Transport in 1998 and the white paper Transport Ten Year Plan 2000 again indicated an acceptance that unrestrained growth in road traffic was neither desirable nor feasible. The worries were threefold: concerns about congestion, concerns about the effect of road traffic on the environment (both natural and built) and concerns that an emphasis on road transport discriminates against vulnerable groups in society such as the poor, the elderly and the disabled.

These documents reiterated the emphasis on integration:

- integration within and between different modes of transport
- integration with the environment
- integration with land use planning
- integration with policies for education, health and wealth creation.

This attempt to reverse decades of underinvestment in the transport system has resulted in a severe shortage of transport planners. It was estimated in 2003 that 2,000 new planners would be required by 2010 to avoid jeopardising the success of the Transport Ten Year Plan .

In 2006, the Transport Planning Society defined the key purpose of transport planning as:

> *to plan, design, deliver, manage and review transport, balancing the needs of society, the economy and the environment.*

The following key roles must be performed by transport planners:

- take account of the social, economic and environmental context of their work
- understand the legal, regulatory policy and resource framework within which they work
- understand and create transport policies, strategies and plans that contribute to meeting social, economic and environmental needs
- design the necessary transport projects, systems and services
- understand the commercial aspects of operating transport systems and services
- know about and apply the relevant tools and techniques
- must be competent in all aspects of management, in particular communications, personal skills and project management.

The UK Treasury recognises and has published guidance on the systematic tendency for project appraisers to be overly optimistic in their initial estimates.

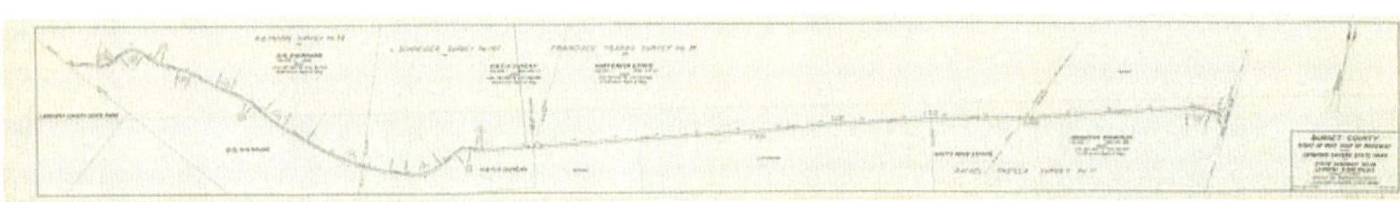

A Civilian Conservation Corps (CCC) map of the planned route of a parkway.
The CCC was actively involved in creating and improving roads throughout rural areas and parks

Transportation planning in the United States is in the midst of a shift similar to that taking place in the United Kingdom, away from the single goal of moving vehicular traffic and towards an approach that takes into consideration the communities and lands through which streets, roads, and highways pass ("the context"). More so, it places a greater emphasis on passenger rail networks, which had been neglected until recently. This new approach, known as Context Sensitive Solutions (CSS), seeks to balance the need to move people efficiently and safely with other desirable outcomes, including historic preservation, environmental sustainability, and the creation of vital public spaces.

The initial guiding principles of CSS came out of the 1998 "Thinking Beyond the Pavement" conference as a means to describe and foster transportation projects that preserve and enhance the natural and built environments, as well as the economic and social assets of the neighborhoods they pass through. CSS principles have since been adopted as guidelines for highway design in federal legislation. Also, in 2003, the Federal Highway Administration announced that under one of its three Vital Few Objectives (Environmental Stewardship and Streamlining) they set the target of achieving CSS integration within all state Departments of Transportation by September 2007.

In recent years, there has been a movement to provide "complete" transportation corridors under the "complete streets" movement. In response to auto-centric design of transportation networks, complete streets encompass all users and modes of transportation in a more equitable manner. The complete streets movement entails many of the CSS principles as well as pedestrian, bicycle and older adult movements to improve transportation in the United States.

These recent pushes for changes to the profession of transportation planning has led to the development of a professional certification program by the Institute of Transportation Engineers, the Professional Transportation Planner in 2007. In response an advanced form of certification - the Advanced Specialty Certification in Transportation Planning was developed by the American Planning Association thereafter in 2011. The Certified Transportation Planner credential is only available for those professional planners (AICP members) who have at a minimum of eight years of transportation planning experience.

The Technical Process

Most regional transport planners employ what is called the rational model of planning. The model views planning as a logical and technical process that uses the analysis of quantitative data to decide how to best invest resources in new and existing transport infrastructure.

Since World War II, this attitude in planning has resulted in the widespread use of travel modelling as a key component of regional transport planning. The models' rise in popularity can also be attributed to a rapid increase in the number of automobiles on the road, widespread suburbanization and a large increase in federal or national government spending upon transport in urban areas. All of these phenomena dominated the planning culture in the late 1940s, 1950s and 1960s. Regional transport planning was needed because increasingly cities weren't just cities anymore, but parts of a complex regional system.

The US process, according to Johnston (2004) and the FHWA and Federal Transit Administration (FTA) (2007), generally follows a pattern which can be divided into three different stages. Over the course of each of three phases, the metropolitan planning organization (MPO) is also supposed to consider air quality and environmental issues, look at planning questions in fiscally constrained way and involve the public. In the first stage, called preanalysis, the MPO considers what problems and issues the region faces and what goals and objectives it can set to help address those issues. During this phase the MPO also collects data on wide variety of regional characteristics, develops a set of different alternatives that will be explored as part of the planning process and creates a list of measurable outcomes that will be used to see whether goals and objectives have been achieved. Johnston notes that many MPOs perform weakly in this area, and though many of these activities seem like the "soft" aspects of planning that aren't really necessary, they're absolutely essential to ensuring that the models used in second phase are accurate and complete .

The second phase is technical analysis. The process involves much technical maneuvering, but basically the development of the models can be broken down as follows. Before beginning, the MPO collects enormous amounts of data. This data can be thought of as falling into two categories: data about the transport system and data about adjacent land use. The best MPOs are constantly collecting this data.

The actual analysis tool used in the US is called the Urban Transportation Modeling System (UTMS), though it is often referred to as the four-step process. As its nickname suggestions, UTMS has four steps: trip generation, trip distribution, mode choice and trip/route assignment. In trip generation, the region is subdivided into a large number of smaller units of analysis called traffic analysis zones (TAZs). Based on the number and characteristics of the households in each zone, a certain number of trips is generated. In the second step, trip distribution, trips are separated out into categories based on their origin and purpose: generally, these categories are home-based work, home-based other and non-home based. In each of three categories, trips are matched to origin and destination zones using the data that has been collected.

In mode choice, trips are assigned to a mode (usually auto or transit) based on what's available in a particular zone, the characteristics of the household within that zone and the cost of the mode for each mode in terms of money and time. Since most trips by bicycle or walking are generally shorter, they are assumed to have stayed within one zone and are not included in the analysis. Finally, in route assignment, trips are assigned to the network. As particular parts of the network are assigned trips, the vehicle speed slows down, so some trips are assigned to alternate routes in such a way that all trip times are equal. This is important because the ultimate goal is system-wide optimization, not optimization for any one individual. The finished product is traffic flows and speeds for each link in the network.

Ideally, these models would include all the different behaviours that are associated with transport, including complex policy questions which are more qualitative in nature. Because of the complexity of transport issues, this is often not possible in practice. This results in models which may estimate future traffic conditions well, but are ultimately based on assumptions made on the part of the planner. Some planners carry out additional sub-system modelling on things like automobile ownership, time of travel, location of land development, location and firms and location of households to help to fill these knowledge gaps, but what are created are nevertheless models, and models always include some level of uncertainty.

The post-analysis phase involves plan evaluation, programme implementation and monitoring of the results. Johnston notes that for evaluation to be meaningful it should be as comprehensive as possible. For example, rather than just looking at decreases in congestion, MPOs should consider economic, equity and environmental issues.

Politics and Transportation Planning

Although a transportation planning process may appear to be a rational process based on standard and objective methodologies, it is often influenced by political processes. Transportation planning is closely interrelated to the public nature of government works projects. As a result, transportation planners play both a technical and a coordinating role. Politicians often have vastly differing perspectives, goals and policy desires. Transportation planners help by providing information to decision makers, such as politicians, in a manner that produces beneficial outcomes. This role is similar to transportation engineers, who are often equally influenced by politics in the technical process of transportation engineering design.

Integration with Urban Planning

Transport isochrone maps can be used in connection with urban planning to evaluate sites.

Urban Transport Planning

Cities play a vital role in promoting economic growth and prosperity of a nation. The development of cities largely depends upon their physical, social, and institutional infrastructure.

Transport demand in most Indian cities has increased substantially, due to increases in population as a result of both natural increase and migration from rural areas and smaller towns. Availability of motorized transport increases household income, and increases in commercial and industrial activities further adds to transport demand. In many cases, demand has outstripped road capacity. Greater congestion and delays are prevalent in Indian cities and indicate the seriousness of transport problems. A high level of pollution is another undesirable feature of overloaded streets. The transport crisis also takes a human toll. Statistics indicate that accidents are a primary cause of accidental deaths in Indian cities. The main reasons for these problems are the prevailing imbalance in modal split, inadequate transport infrastructure, and its suboptimal use. Public transport systems in metropolitan cities have not been able to keep pace with the rapid and substantial increases in travel demand over the past few decades. Bus services in particular have deteriorated, and their relative output has been further reduced as passengers have turned to personalized modes and intermediate public transport.

Individual cities cannot afford to cater only to private cars and two-wheelers. There must be a general recognition that without public transport cities would be even less viable. There is a need to promote public transport instead of personal vehicles. This requires both an increase in quantity as well as quality of public transport and effective use of demand as well as supply-side management measures. People should also be encouraged to use non-motorized transport and investments may be made to make it safer. Cities are the major contributors to economic growth, and movements within cities are crucial for improved quality of life.

Transportation Planning Processes

Urban transportation planning is the process that leads to decisions on transportation policies and programs. In this process, planners develop information about the impacts of implementing alternative courses of action involving transportation services, such as new highways, introduction of new modes of public transport etc, or parking restrictions. The fundamental objective of transportation is to provide efficient and safe levels of mobility required to support a wide spectrum of human needs for a heterogeneous variety of societal groups. Because these needs, goals, and objectives are continuously changing, transportation planning is also an ever-evolving process. The important steps of the transportation planning process are as given below:

Step 1: Forecasting target year population and economic growth for the subject metropolitan area.

Step 2: allocation of land use and socio – economic projections individual analysis zones according to land availability, local zoning and related public policies.

Step 3: specification of alternative transportation plans partly based on the result of Step 1 and Step 2.

Step 4: calculation of the capital and maintenance costs of each alternative plan.

Step 5: application of calibrated demand – forecasting models to predict target year equilibrium flows expected to use each alternative, given the land use and socio – economic projection of Step 2 and the characteristics of the transportation alternatives (Step 3).

Step 6: conversion of equilibrium flows to direct user benefits, such as savings in travel time and travel cost attributable to the proposed plan.

Step 7: comparative evaluation and selection of the best of the alternatives analysed based on estimated costs (Step 3) and benefits (Step 6).

This information is used to help decision-makers (elected officials or their representatives) in their selection of transportation policies and programs.

Problem Definition

Problem is defined by defining the objective/goal of planning. The objective could be

i. To develop a transport system that caters to all sections of society thereby promoting inclusive development.

ii. To develop a transportation system that is primarily sustainable in nature.

iii. To minimize system costs and provide self operation etc.

Identification of Need

- Inventory of existing state of system and a forecast of future condition (demand).
- Interpolation of objective.

4.1.3 Solution Generation and Analysis

Based on problem definition, transportation planners have to identify various alternatives and make choice about-

- Various modes like roadway (DA/PT), railway, waterways or airways.
- The technological aspect such as high speed train, raised monorails underground transit system, driver information system.
- Traffic engineering aspect such as changing or improving flow pattern by making certain road one way, reducing delay on arterial street by improving signalization or grade separated intersection, disallowing certain movement at intersection
- Regulatory aspect such as reserving land for only high occupancy vehicles, disallowing high polluting vehicles, imposing speed limit.

For example, to travel from Kanpur to Delhi following alternatives can be worked out.

- o Facilitate train leaving Kanpur at 6:30 am and reaching Delhi at 10:30 am and again leaving Delhi at 5:30pm and reaching Kanpur at 10:30pm.
- o Developing existing airport at Kanpur for small commercial flight
- o Improve the existing road facilities providing operating speed 120 kmph.

Public Transport Planning

In the final years of the 20th century, public transit and its planning and development have faced continuing changes in current ways of doing business. At the same time, there has been is increasing recognition that current trends and their consequences (decentralized growth and development, wasteful environmental consumption, and dramatically increased traffic congestion) are not supportive of the lifestyles or quality of life that most Americans aspire to. Efforts to improve community planning and to foster a multimodal transportation system are gaining support. A policy debate has been generated at the local and state levels initially but is now likely to become part of the political and policy agenda in Washington, D.C., as well. The renewed acceptance of the merits of planning concepts, alternative development patterns to avoid sprawl, and the multiple roles and benefits of public transit bode well for the transit industry and for the agenda of the Committee on Public Transportation Planning and Development and offers new opportunities as we look toward the 21st century.

Current Approaches

Public transit and planning and development are undergoing major changes prompted by both external factors and policy-directed responses.

Public Transit

Public transit has made tremendous strides to adapt to a changing environment (or setting) that is, in most cases, contrary to its best interests or even survival. Through ongoing and deliberate self-examination and promotion, the industry is striving to

- Incorporate the latest technological advances (intelligent transportation systems) in day-to-day operations and fare collection practices;

- Expand the mobility manager concept to provide more flexible and adaptable responses to the realities of changing local and regional travel markets and needs;
- Focus transit employees on the need to foster a customer-first or hospitality industry approach to the riding public;
- Consider new institutional arrangements, including public-private cooperative ventures, to provide, manage, and implement service and project initiatives; and
- Implement expanded research and training programs that explore new approaches and incorporate them into current practice

The bottom line of these multitrack initiatives is that the public transit industry has recognized that it must adapt or die given the realities of the competitive marketplace, technological advances, and the diverse and sometimes hostile setting for the provision of service.

The planning and development fields are also changing and adapting to external conditions:

- The renewed focus on neotraditional development patterns, while sometimes viewed as "old wine in a new bottle," is beginning to attract more attention from consumers and the development community, as well as the media.
- Changing demographics and rising prosperity are leading changes in the quality of community development and the attractiveness of a variety of settings, including a return to urban living.
- The central role of public transit, particularly rail transit, in shaping transit-oriented development is gaining strength (even though the empirical evidence is still mixed).
- Efforts to foster multimunicipal planning and local-county cooperation in fragmented jurisdictions are gaining support, although the home-rule mentality is likely to continue to dominate. One of the biggest successes is the growing movement of state-level initiatives that offer a more consistent policy and coordinated infrastructure investment overlay of county and local government planning activities.
- Experiments with such techniques as urban growth boundaries and concurrency between development approval and infrastructure availability are being implemented or considered by localities.
- Outreach techniques and public involvement programs to foster participatory planning are gaining acceptance.

In general, current prospects for planning and development issues, particularly as they relate to developing the settings for public transit to thrive and prosper, are favorable.

Prospects

Macro Issues

Central to any discussion of the future is recognition of various macro issues that will affect everything else that occurs in the 21st century. Issues of this type include

- The effect of technology on daily life and business;

- The increasingly global and interconnected economy and financial structure;
- The aging of the population, both nationally and internationally;
- An increasing disparity between the have and have-not nations, populations, and elements of society;
- Conflicts between sustainable practices and continued consumption of resources at existing rates; and
- Changing notions of nations, communities, and individuals.

These macro issues can bring about a greater concern for planning and the consequences of development decisions and can promote the market for public transit (for example, sustainability versus consumption and the increasingly elderly population). However, they can also detract from planning's core goals (such as compactness and coordinated infrastructure) and the provision of public transit (for example, the increasing digital communication connections reduce the need for compact development and the traditional role of cities and urban places as central gathering places for commerce).

Local Issues

Local policy and project-level responses, for both planning and public transit, will still require assessments of local trends, issues, and goals, in cooperation with affected community residents and interests.

Local issue assessments will need to examine such areas as

- Changing demographics,
- Changing businesses,
- Multiple (transit service) markets,
- Tcchnological effects,
- Community attitudes,
- Financial capacity,
- Institutional structure,
- Environmental attributes, and
- Unmet needs (social service and physical).

Public transit can continue to play a role in responding to these issues and, where possible, can work toward bending the trends through leadership and example. As public transportation planners, we face a continuing challenge to refine our ability to model the predicted interactions between land uses and transportation demand and supply. Related to that challenge is the development of better techniques to manage, recall, and evaluate information as the new information technology gives us more powerful tools to acquire and store data. Expert systems and other approaches to artificial intelligence could be harnessed, for example, to provide fast recall of examples of solutions to transportation problems in analogous settings. Reasoning by analogy can provide both alternatives to be evaluated and information useful to model calibration and development of other forecasting tools.

In general, those involved in the planning and development aspects of the public transit field should be optimistic about the prospects for public transit and the continued need for a strong role for planning. At the same time, continued monitoring of evolving global and local issues will be essential to ensure that planning for public transit is in step with changing community needs, including those of both residents and business enterprises. Continued searches for innovative practices, particularly those that maximize new technology in cost-effective ways, will be critical for transit's success. The committee's agenda, as embodied in its strategic plan, can help the Transportation Research Board to participate in the 21st century challenges ahead.

Traffic Analysis Zone

Traffic analysis zones are boundaries that contain socioeconomic data that is used as the foundation for trip-making in the travel model. The TAZ layer is formatted as a polygon layer in TransCAD's GIS structure. The TAZs are attached to the networks using zone centroids and centroid connectors that allow travelers access to the transportation system by simulating local and neighborhood streets.

TAZs are ideally but not always sized and shaped to provide a relatively homogeneous amount and type of activity within each zone. TAZ delineations traditionally follow the natural and man-made boundaries that tend to segregate different land uses. These boundaries include water features, roads, railroads, and other lines that form logical boundaries. Jurisdictional and census boundaries often do not make for good TAZ definitions because they can be arbitrary in relation to the needs of the model; but they are usually desirable for data development and reporting functions.

The definition of traffic analysis zones has implications throughout the travel model. For roadway model components, traffic analysis zone resolution affects the amount of precision that can be achieved when loading vehicles onto the collector and arterial roadway network. This precision is obtained by increasing detail in the roadway network, TAZ structure, and socioeconomic data. The desire for increased detail must however be balanced with increased computer processing times and the ability to develop and maintain the data at the increased level of detail.

For transit model components, the size of traffic analysis zones affects the accuracy of transit path-building, particularly the walk to transit component. Algorithms used in the mode choice model will need to be designed based on the TAZ structure.

Access Management

Access management is the systematic control of the location, spacing, design, and operation of driveways, median openings, interchanges, and street connections to a highway. It also involves highway design applications, such as median treatments and auxiliary lanes, and the appropriate spacing of traffic signals.

The purpose of access management is to provide vehicular access to land development in a manner that preserves the safety and efficiency of the transportation system. The practice of access management extends the concept of access design and location control to all highways—not just controlled-access highways or freeways.

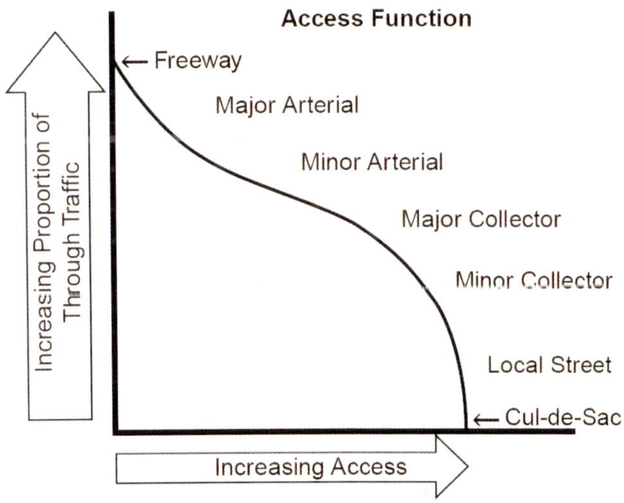

From a planning viewpoint, access management is a systematic way to carry out the highway functional hierarchy implicit in state, regional, and local transportation plans. Highways are classified by function on the basis of the priority given to land access versus through-traffic movement.

Access management is particularly important along arterials and other primary roads that are expected to provide safe and efficient movement of traffic, as well as access to property. However, access management is still necessary on lower-level roads, such as minor collectors and residential streets, to address safety considerations, such as sight distance and corner clearance.

Principles of Access Management

Access management programs seek to limit and consolidate access along major roadways, while promoting a supporting street system and unified access and circulation systems for development. The result is a roadway that functions safely and efficiently for its useful life, and a more attractive corridor. The goals of access management are accomplished by applying the following principles:

1. Provide a Specialized Roadway System: Different types of roadways serve different functions. It is important to design and manage roadways according to the primary functions that they are expected to serve.

2. Limit Direct Access to Major Roadways: Roadways that serve higher volumes of regional through traffic need more access control to preserve their traffic function. Frequent and direct property access is more compatible with the function of local and collector roadways.

3. Promote Intersection Hierarchy: An efficient transportation network provides appropriate transitions from one classification of roadway to another. For example, freeways connect to arterials through an interchange that is designed for the transition. Extending this concept to other roadways results in a series of intersection types that range from the junction of two major arterial roadways, to a residential driveway connecting to a local street.

4. Locate Signals to Favor Through Movements: Long, uniform spacing of intersections and signals on major roadways enhances the ability to coordinate signals and to ensure continuous movement of traffic at the desired speed. Failure to carefully locate access connections or median openings that later become signalized, can cause substantial increases in arterial travel times. In addition, poor signal placement may lead to delays that cannot be overcome by computerized signal timing systems.

5. Preserve the Functional Area of Intersections and Interchanges: The functional area of an intersection or interchange is the area that is critical to its safe and efficient operation. This is the area where motorists are responding to the intersection or interchange, decelerating, and maneuvering into the appropriate lane to stop or complete a turn. Access connections too close to intersections or interchange ramps can cause serious traffic conflicts that result in crashes and congestion.

6. Limit the Number of Conflict Points: Drivers make more mistakes and are more likely to have collisions when they are presented with the complex driving situations created by numerous conflict points. Conversely, simplifying the driving task contributes to improved traffic operations and fewer collisions. A less complex driving environment is accomplished by limiting the number and type of conflicts between vehicles, vehicles and pedestrians, and vehicles and bicyclists.

7. Separate Conflict Areas: Drivers need sufficient time to address one set of potential conflicts before facing another. The necessary spacing between conflict areas increases as travel speed increases, to provide drivers adequate perception and reaction time. Separating conflict areas helps to simplify the driving task and contributes to improved traffic operations and safety.

8. Remove Turning Vehicles from Through Traffic Lanes: Turning lanes allow drivers to decelerate gradually out of the through lane and wait in a protected area for an opportunity to complete a turn. This reduces the severity and duration of conflict between turning vehicles and through traffic and improves the safety and efficiency of roadway intersections.

9. Use Nontraversable Medians to Manage Left-Turn Movements: Medians channel turning movements on major roadways to controlled locations. Research has shown that the majority of access-related crashes involve left turns. Therefore, nontraversable medians and other techniques that minimize left turns or reduce the driver workload can be especially effective in improving roadway safety.

10. Provide a Supporting Street and Circulation System: Well-planned communities provide a supporting network of local and collector streets to accommodate development, as well as unified property access and circulation systems. Interconnected street and circulation systems support alternative modes of transportation and provide alternative routes for bicyclists, pedestrians, and drivers. Alternatively, commercial strip development with separate driveways for each business forces even short trips onto arterial roadways, thereby reducing safety and impeding mobility.

The People Benefitted from Access Management

Motorists

- Face fewer decision points and traffic conflicts, which simplifies driving and increases driver safety
- Experience fewer traffic delays and arrive more quickly at their destinations

Cyclists

- Face fewer decision points and conflicts with traffic, which simplifies cycling and increases safety for cyclists
- Benefit from more predictable motorist travel patterns
- Can select alternative travel routes as local supporting highway systems are developed

Pedestrians

- Face fewer and less frequent access points where motorists enter and exit the roadway making it safer to walk along major roadways
- Can use medians as a refuge when crossing several lanes of traffic

Transit Riders

- Experience reduced delay and reduced travel times
- Benefit from a safer walking environment
- Experience more convenient access to transit stops as connectivity of streets, sidewalks, and pedestrian ways is improved

Business Persons

- Are served by a more efficient roadway system that captures a broader market area
- Benefit from stable property values due to a well-managed roadway corridor
- Experience a more predictable and consistent development environment

The freight industry benefits from reduced delay and increased safety, which results in lower transportation costs and shorter delivery times

Government Agencies

- Benefit from the lower cost of delivering an efficient and safe transportation system
- Benefit from improved internal and intergovernmental coordination
- Are more effective in accomplishing their transportation objectives

Communities

- Receive a safer transportation system
- Benefit from less need for road widening, which causes displacement of businesses, homes, and communities
- Benefit from more attractive roadway corridors
- Help protect and preserve their investment in transportation facilities and may reduce capital improvement costs for new or reconstructed roadways

The Effects of the Mismanagement of Roadway Access

Without access management, the function and character of major highway corridors can deteriorate rapidly. Failure to manage access is associated with the following adverse social, economic, and environmental impacts:

- An increase in vehicular crashes
- More collisions involving pedestrians and cyclists
- Accelerated reduction in highway efficiency
- Unsightly commercial strip development
- Degradation of scenic landscapes

- More cut-through traffic in residential areas due to overburdened arterials
- Homes and businesses adversely affected by a continuous cycle of widening roads
- Increased commute times, fuel consumption, and vehicle emissions as numerous driveways and traffic signals increase congestion and delays along major highways

Not only is inadequate access management costly for government agencies and the public, but it also adversely affects corridor businesses. Over time, closely spaced and poorly designed driveways make it more difficult for customers to enter and exit businesses safely. Access to corner businesses may be blocked by queuing traffic. Newer businesses seek out locations that are more attractive and have fewer access and congestion problems. Customers begin to patronize businesses with safer, more convenient access and avoid businesses in areas of poor access design. Gradually the older developed areas begin to deteriorate because of access and aesthetic problems, and investment moves to newer, better-managed corridors.

Annual Average Daily Traffic

Transportation planners use a variety of traffic volume statistics to monitor and plan for traffic on roadways. Two of the most important statistics used are ADT, or average daily traffic, and AADT, or average annual daily traffic.

Annual Average Daily Traffic (AADT) is the estimated mean daily traffic volume. For continuous sites, calculated by summing the Annual Average Days of the Week and dividing by seven. For short-count sites, estimated by factoring a short count using seasonal and day-of-week adjustment factors. Commercial Annual Average Daily Traffic (CADT) is the estimated mean daily traffic volume for commercial vehicles. Values are calculated using the same procedures as AADT.

Automated traffic counters on major roadways, installed by state departments of transportation, provide continuous counting of vehicle traffic. Temporary automated counters -- including road tubes or video cameras -- collect traffic data over a shorter duration, usually less than a week.

Shorter-term traffic counts -- weekly or monthly -- estimate annual traffic demand when complete annual counts are not available. Adjustments to actual traffic counts to reflect daily, weekly and seasonal variations of traffic on the roadway under study provide an estimate of annual traffic.

Analysts use average daily traffic estimates to monitor the growth in traffic on a roadway from year to year and for programming and funding of major improvements, such as roadway widening. Average daily traffic is also useful for analyzing the rate of traffic accidents on a roadway.

Uses

Highway 401 in Ontario, Canada, has an AADT of over 400,000 in some sections of Toronto.

One of the most important uses of AADT is for determining funding for the maintenance and improvement of highways.

In the United States the amount of federal funding a state will receive is related to the total traffic measured across its highway network. Each year on June 15, every state in the United States submits a Highway Performance Monitoring System HPMS report. The HPMS report contains various information regarding the road segments in the state based on a sample (not all of the road segments) of the road segments. In the report, the AADT is converted to vehicle miles traveled (VMT). VMT is the AADT multiplied by the length of the road segment. To determine the amount of traffic a state has, the AADT cannot be summed for all road segments since an AADT is a rate. The VMT is summed and is used as an indicator of the amount of traffic a state has. For federal-funding, formulas are applied to include the VMT and other highway statistics.

In the United Kingdom AADT is one of a number of measures of traffic used by local highway authorities, Highways England and the Department for Transport to forecast maintenance needs and expenditure.

Data Collection

To measure AADT on individual road segments, traffic data is collected either by an automated traffic counter or hiring an observer to record traffic. There are two different techniques of measuring the AADTs for road segments. One technique is called continuous count data collection method. This is where sensors are permanently embedded into a road and traffic data is measured all 365 days. The AADT would be the sum of the total traffic for the entire year divided by 365 days. There is a problem with calculating the AADT with this method. The continuous count equipment is not operating for the full 365 days due to being shut down for maintenance or repair. Because of this, seasonal or day-of-week biases might skew the calculated AADT. In 1992, AASHTO released the AASHTO Guidelines for Traffic Data Programs, which identified a way to produce an AADT without seasonal or day-of-week biases by creating an "average of averages." For every month and day-of-week, a Monthly Average Day of Week (MADW) is calculated (84 per year). Each day-of-week's MADW is then calculated across months to calculate an Annual Average Day of Week (AADW) (7 per year). Finally, the AADWs are averaged to calculate an AADT. The United States Federal Highway Administration (FHWA) has adopted this method as the preferred method in the [FHWA Traffic Monitoring Guide].

A traffic counter on BIA Road J-9 in the United States

While providing the most accurate AADT, installing and maintaining continuous count stations method is costly. Most agencies are only able to monitor a very small percentage of the roadway using this method. Most AADTs are generated using short-term data collection methods sometimes known as the coverage count data collection method. Traffic is collected with portable sensors that are attached to the road and record traffic data typically for 2 – 14 days. These are typically pneumatic road tubes although other more expensive technology such as radar, laser, or sonar exist. After recording the traffic data, the traffic counts on the same road segment are taken again in another three years. FHWA Traffic Monitoring Guide recommends performing a short count on a road segment at a minimum of every three years. There are many methods used to calculate an AADT from a short-term count, but most methods attempt to remove seasonal and day-of-week biases during the collection period by applying factors created from associated continuous counters. Short counts are taken either by state agencies, local government, or contractors.

For the years when a traffic count is not recorded, the AADT is often estimated by applying a factor called the Growth Factor. Growth Factors are statistically determined from historical data of the road segment. If there is no historical data, Growth Factors from similar road segments are used.

Annual Average Weekday Traffic (AAWT)

Annual average weekday traffic (AAWT) is similar to AADT but only includes Monday to Friday data. Public holidays are often excluded from the AAWT calculation.

Average Summer Daily Traffic

Average summer daily traffic (abbreviated to *ASDT*) is a similar measure to the annual average daily traffic. Data collecting methods of the two are exactly the same, however the ASDT data is collected during summer only. The measure is useful in areas where there are significant seasonal traffic volumes carried by a given road.

Average Daily Traffic

Average daily traffic or ADT, and sometimes also mean daily traffic, is the average number of vehicles two-way passing a specific point in a 24-hour period, normally measured throughout a year. ADT is not as highly referred to as the engineering standard of AADT which is the standard measurement for vehicle traffic load on a section of road, and the basis for most decisions regarding transport planning, or to the environmental hazards of pollution related to road transport.

Bicycle Transportation Planning and Engineering

More bicyclists and more safety concerns creates need for bicycle transportation engineers

More cyclists on the road mean more bikes on the road, the same road that is home to daily car and public transportation commuters. Sharing the road is necessary, but has also shown to be a cause for concern.

According to the Center for Disease Control, adult bicycling-related fatalities have increased three-fold for those between 35 and 74 years of age. Florida holds the distinction of being perhaps the most dangerous state for cyclists, with bicycle-related accidents being two times more likely to be fatal than the rest of the country.

Reasons for this increase can be traced to several categories:

- The growing number of bicyclists.
- Improper road design due to engineering oversight.
- Lack of current traffic law enforcement.
- Driver and cyclist behaviors.
- The lack of helmet use.
- The growing amount of traffic and cycling volume using the same roads at the same time.

Safe streets initiative challenges cities way of thinking

With the trend towards bicycling and commuter bicycling growing yearly, many cities are realizing the need for safe facilities and safer roads. And through initiatives such as the Safer People, Safer Streets, from U.S. Transportation Secretary Anthony Foxx, safety has come to the forefront in the form of a Mayor's Challenge, being directed towards every state.

Foxx issued the challenge to address non-motorized safety issues, and to help communities create safer, better-connected bicycle and walking networks throughout their cities.

He has challenged individual transportation agencies to go beyond their minimum requirements and for them to be proactive in providing the places and facilities that encourage bicycle use across all age groups.

This initiative has led to strategic, detailed studies centered on bicyclist-friendly options. While raising the importance of safety, it has also led to an increased need for new personnel to take on the responsibility of studying the balance of transportation options within cities, and through more informed decision making and smarter investments, making sure that those balances are considered in future traffic projects.

Commercial outdoor bike racks for multiple bikes

While many cities now have staff dedicated to bicycle and pedestrian safety, there are also many outside, individual, and special interest groups promoting their plans and projects as well. An organized effort was lacking.

Combining Efforts and Organizing Initiatives

There was clearly a need for someone to oversee all these separate yet related initiatives while keeping focus on an overall, master plan.

Sprouting from this lack of long term, cohesive planning, cities started appointing *Bike Transportation Engineers*, or Bike and Pedestrian Coordinators to be the person in charge of combining all transportation-related efforts in a unified way.

While many cities were already looking into this trend, others, like St. Louis have just now become serious about the matter. They had made significant investments in bicycle infrastructure over the past several years, but it seemed to be in a piecemeal sort of way.

Rather than continue on in this manner, the city felt they needed an organizer to oversee the process, and work not for a few, but for all those that would be involved and affected.

So this past fall, the new position of Bike and Pedestrian Coordinator was created. It was initially presented only as "a professional engineering position charged with creating a more sustainable, safer, more inviting city for residents and visitors".

A short time later, Jamie Wilson, former president of a traffic engineering firm was hired, as St. Louis Mayor Francis Slay further explained, "To help make St. Louis a more bikeable, walkable place to live, work and explore by considering everyone, not just drivers, when it comes to the design and maintenance of streets".

The position is to use a targeted, data driven approach to funnel investment resources for improvements where they are most needed. You've seen those cables that lay across the road to count traffic? Now that same principle is used to also track bicyclist traffic, showing the most used routes.

Wilson is now tasked with coordinating a master plan between all involved and affected organizations so as to educate drivers, bicyclists and pedestrians in a way that will encourage the use and implementation of improvements such as bike lanes (both dedicated and shared), appropriate signals and signage, public bike racks and parking corrals, and ADA access.

Safety is the Overall Mission

Wilson accepted the position because he believes, "It's a quality of life issue. The cyclists are here, right now, and having traffic conflicts and safety issues on a daily basis. We need to react, teach, and learn to share our roads. We needed someone running point for all the separate but related projects that go on within a city, not only internally but for projects included in private development as well.

"If a consultant comes in with a proposal, we need to make sure that all the modes of transportation are taken into consideration. All projects must fit into our bigger picture in an organized way. Safety is the overall mission".

Wave commercial bike racks create more parking for the public

While safety is most certainly important, Bike Engineers can also have an effect on other areas, like the tourist industry and visitors who like to ride bikes.

They can lead the way to build programs related to a city's unique history and culture. For example, in St. Louis, through a coordinated effort between Jazz St. Louise, a jazz education and outreach group, and Trailnet, a non-profit that promotes active, healthy living within communities, there exists a hosted bike tour of the city's rich musical history.

Safe biking environments, roads and bicycle parking racks provide current residents, visitors, and tourists a way to explore the city on different levels, with a more unique and individualized perspective.

Transportation Forecasting

Transportation forecasting is the process of estimating the number of people or vehicles that will use a specific transportation facility in the future. Transportation forecasts can be utilized in a variety of different situations and with different modes of transport, from estimating traffic volumes on a specific segment of road or highway, to estimating ships in a port, or passenger volumes on a city's buses.

Traffic forecasts help to explain what the needs of the future might be and provide benchmarks for proper design and efficient transportation system operation. Example applications of forecast information include:

- Development of infrastructure capacity and design calculations (e.g., the operations of an existing or proposed roadway or bridge, or the thickness or type of roadway pavements)
- Estimation of the financial and/or social viability of projects (e.g., developing benefit-cost analyses and/or social impact assessments)
- Calculation of environmental impacts, such as air and noise pollution

Four-step Models

Within the rational planning framework, transportation forecasts have traditionally followed the sequential four-step model or urban transportation planning (UTP) procedure, first implemented on mainframe computers in the 1950s at the Detroit Metropolitan Area Traffic Study and Chicago Area Transportation Study (CATS).

Land-use forecasting starts the process. Typically, forecasts are made for the region as a whole, e.g., of population growth. Such forecasts provide control totals for the local land use analysis. Typically, the region is divided into zones and by trend or regression analysis, the population and employment are determined for each.

The four steps of the classical urban transportation planning system model are:

- Trip generation determines the frequency of origins or destinations of trips in each zone by trip purpose, as a function of land uses and household demographics, and other socio-economic factors.
- Trip distribution matches origins with destinations, often using a gravity model function, equivalent to an entropy maximizing model. Older models include the Fratar or Furness method, a type of iterative proportional fitting.
- Mode choice computes the proportion of trips between each origin and destination that use a particular transportation mode (this modal model may be of the logit form).
- Route assignment allocates trips between an origin and destination by a particular mode to a route. Often (for highway route assignment) Wardrop's principle of user equilibrium is applied (equivalent to a Nash equilibrium), wherein each driver (or group) chooses the shortest (travel time) path, subject to every other driver doing the same. The difficulty is that travel times are a function of demand, while demand is a function of travel time, the so-called bi-level problem. Another approach is to use the Stackelberg competition model, where users ("followers") respond to the actions of a "leader", in this case for example a traffic manager. This leader anticipates on the response of the followers.

After the classical model, there is an evaluation according to an agreed set of decision criteria and parameters. A typical criterion is cost–benefit analysis. Such analysis might be applied after the network assignment model identifies needed capacity: is such capacity worthwhile? In addition to identifying the forecasting and decision steps as additional steps in the process, it is important to note that forecasting and decision-making permeate each step in the UTP process. Planning deals with the future, and it is forecasting dependent.

Activity-based Models

Activity-based models are another class of models that predict for individuals where and when specific activities (e.g. work, leisure, shopping, ...) are conducted.

The major premise behind activity-based models is that travel demand is derived from activities that people need or wish to perform, with travel decisions forming part of the scheduling decisions. Travel is then seen as just one of the attributes of a system. The travel model is therefore set within the context of an agenda, as a component of an activity scheduling decision.

Activity-based models offer other possibilities than four-step models, e.g. to model environmental issues such as emissions and exposure to air pollution. Although their obvious advantages for environmental purposes were recognized by Shiftan almost a decade ago, applications to exposure models remain scarce. Activity-based models have recently been used to predict emissions and air quality. They can also provide a better total estimate of exposure while also enabling the disaggregation of individual exposure over activities. They can therefore be used to reduce exposure misclassification and establish relationships between health impacts and air quality more precisely. Policy makers can use activity-based models to devise strategies that reduce exposure by changing time activity patterns or that target specific groups in the population.

Integrated Transport - Land use Models

These models are intended to forecast the effect of changes in the transport network and operations over the future location of activities, and then forecast the effect of these new locations over the transport demand.

Per-driver Models

As data science and big data technologies become available to transport modelling, research is moving towards modelling and predicting behaviours of individual drivers in whole cities at the individual level. This will involve understanding individual drivers' origins and destinations as well as their utility functions. This may be done by fusing per-driver data collected on road networks, such as my ANPR cameras, with other data on individuals, such as data from their social network profiles, store card purchase data, and search engine history. This will lead to more accurate predictions, enhanced ability to control traffic for customized prioritization of particular drivers, but also to ethical concerns as local and national governments use more data about identifiable individuals. While the integration of such partially personal data is tempting, there are considerable privacy concerns over the possibilities, related to the criticisms of mass surveillance.

Precursor Steps

Although not identified as steps in the UTP process, a lot of data gathering is involved in the UTP analysis process. Census and land use data are obtained, along with home interview surveys and journey surveys. Home interview surveys, land use data, and special trip attraction surveys provide the information on which the UTP analysis tools are exercised.

Data collection, management, and processing; model estimation; and use of models to yield plans are much used techniques in the UTP process. In the early days, in the USA, census data was augmented that with data collection methods that had been developed by the Bureau of Public Roads (a predecessor of the Federal Highway Administration): traffic counting procedures, cordon "where are you coming from and where are you going" counts, and home interview techniques. Protocols for coding networks and the notion of analysis or traffic zones emerged at the CATS.

Model estimation used existing techniques, and plans were developed using whatever models had been developed in a study. The main difference between now and then is the development of some analytic resources specific to transportation planning, in addition to the BPR data acquisition techniques used in the early days.

Critique

The sequential and aggregate nature of transportation forecasting has come under much criticism. While improvements have been made, in particular giving an activity-base to travel demand, much remains to be done. In the 1990s, most federal investment in model research went to the Transims project at Los Alamos National Laboratory, developed by physicists. While the use of supercomputers and the detailed simulations may be an improvement on practice, they have yet to be shown to be better (more accurate) than conventional models. A commercial version was spun off to IBM, and an open source version is also being actively maintained as TRANSIMS Open-Source.

A 2009 Government Accountability Office report noted that federal review of transportation modeling focused more on process requirements (for example, did the public have adequate opportunity to comment?) than on transportation outcomes (such as reducing travel times, or keeping pollutant or greenhouse gas emissions within national standards).

One of the major oversights in the use of transportation models in practice is the absence of any feedback from transportation models on land use. Highways and transit investments not only respond to land use, they shape it as well.

Benefits of Transportation Forecasting

Transportation forecasting solutions ensure that logistics, manufacturing, and supply chain are executing to the same plan. Transportation and warehouse planners have visibility into promotions so they can plan for spikes in demand, reducing their reliance on the spot market and enabling more accurate staffing.

Although premiums are not high now, transportation is a cyclical business and rates will rise as capacity has dropped and business is improving. Manufacturers can increase intermodal shipments and decrease deadhead miles with the ability to plan for increased demand weeks in advance of actual orders. Switching to intermodal shipments reduces carbon emissions and fuel consumption, decreasing costs and improving sustainability efforts. Visibility into future transportation requirements enables manufacturers to identify capacity issues in advance, giving logistics the time to resolve them. Visibility into future shipments and receipts improves warehouse staffing. Separating truck procurement from order creation both lowers costs and improves responsiveness. Forecasts that match current manufacturing and distribution capabilities and strategies generate more effective strategic sourcing and give manufacturers the opportunity to collaborate with carriers and retailers, benefitting manufacturers, retailers, carriers, and consumers.

Transportation forecasts can be integrated with Transportation and Warehouse Management Systems to further improve efficiency. Manufacturers can cut costs, improve service, and increase sustainability by forecasting transportation requirements rather than just reacting to orders.

References

- Robinson, Darren, ed. (Nov 12, 2012). "6". Computer Modelling for Sustainable Urban Design: Physical Principles, Methods and Applications. Routledge. p. 157. ISBN 9781136539350. Retrieved 6 October 2017
- "Technical Guidance on Accessibility Planning in Local Transport Plans" (PDF). Local Transport Planning Network. Archived from the original (PDF) on 20 November 2007. Retrieved 26 March2012
- Shiftan Y. (2000). "The advantage of activity-based modelling for air-quality purposes: theory vs practice and future needs". Innovation. 13 (1): 95–110. doi:10.1080/135116100111685
- Michael Meyer, Eric J. Miller. Urban Transportation Planning, McGraw-Hill, 2nd edition, 2000. ISBN 0-07-242332-3
- Barker, Kate (December 2006). "Barker Review of Land Use Planning" (PDF). Barker Review of Land Use Planning. Retrieved March 26, 2012
- Int Panis L (2010). "New Directions: Air pollution epidemiology can benefit from activity-based models". Atmospheric Environment. 44: 1003–1004. doi:10.1016/j.atmosenv.2009.10.047

3

Highway Engineering

Highway engineering is a sub-field of civil engineering that is concerned with the planning, design, construction and management of roads, tunnels and bridges for the efficient transportation of people and goods. This chapter has been carefully written to provide an easy understanding of the varied aspects of highway engineering like geometric design of roads or highways, road surface and road infrastructure.

Road transport is one of the most common mode of transport. Roads in the form of trackways, human pathways etc. were used even from the pre-historic times. Since then many experiments were going on to make the riding safe and comfort. Thus road construction became an inseparable part of many civilizations and empires.

The road pavements are generally constructed on small embankments, slightly above the general ground level wherever possible, in order to avoid the difficult drainage and maintenance problems. The term road or roadway thus constructed is therefore termed 'highway' and the science and technology dealing with Road Engineering or Traffic Engineering is generally called 'Highway Engineering'.

In nutshell it may be said that the highway engineering is a branch of Traffic engineering that deals with the various phase like development, planning, alignment, highway geometric design and location, highway traffic operation and its control, materials, pavement design, construction and maintenance, economic consideration, finance and administration.

Planning and Development

Highway planning involves the estimation of current and future traffic volumes on a road network. Highway engineers strive to predict and analyze all possible civil impacts of highway systems. Some considerations are the adverse effects on the environment, such as noise pollution, air pollution, water pollution, and other ecological impacts.

Financing

Developed countries are constantly faced with high maintenance cost of aging transportation highways. The growth of the motor vehicle industry and accompanying economic growth has generated a demand for safer, better performing, less congested highways. The growth of commerce, educational institutions, housing, and defense have largely drawn from government budgets in the past, making the financing of public highways a challenge.

The multipurpose characteristics of highways, economic environment, and the advances in highway pricing technology are constantly changing. Therefore, the approaches to highway financing, management, and maintenance are constantly changing as well.

Environmental Impact Assessment

The economic growth of a community is dependent upon highway development to enhance mobility. However, improperly planned, designed, constructed, and maintained highways can disrupt the social and economic characteristics of any size community. Common adverse impacts to highway development include damage of habitat and bio-diversity, creation of air and water pollution, noise/vibration generation, damage of natural landscape, and the destruction of a community's social and cultural structure. Highway infrastructure must be constructed and maintained to high qualities and standards.

There are three key steps for integrating environmental considerations into the planning, scheduling, construction, and maintenance of highways. This process is known as an Environmental Impact Assessment, or EIA, as it systematically deals with the following elements:

1. Identification of the full range of possible impacts on the natural and socio-economic environment
2. Evaluation and quantification of these impacts
3. Formulation of measures to avoid, mitigate, and compensate for the anticipated impacts.

Highway Safety

Highway systems generate the highest price in human injury and death, as nearly 50 million persons are injured in traffic accidents every year, not including the 1.2 million deaths. Road traffic injury is the single leading cause of unintentional death in the first five decades of human life.

Management of safety is a systematic process that strives to reduce the occurrence and severity of traffic accidents. The man/machine interaction with road traffic systems is unstable and poses a challenge to highway safety management. The key for increasing the safety of highway systems is to design, build, and maintain them to be far more tolerant of the average range of this man/machine interaction with highways. Technological advancements in highway engineering have improved the design, construction, and maintenance methods used over the years. These advancements have allowed for newer highway safety innovations.

By ensuring that all situations and opportunities are identified, considered, and implemented as appropriate, they can be evaluated in every phase of highway planning, design, construction, maintenance, and operation to increase the safety of our highway systems.

Design

The most appropriate location, alignment, and shape of a highway are selected during the design stage. Highway design involves the consideration of three major factors (human, vehicular, and roadway) and how these factors interact to provide a safe highway. Human factors include reaction time for braking and steering, visual acuity for traffic signs and signals, and car-following behaviour. Vehicle considerations include vehicle size and dynamics that are essential for determining lane width and maximum slopes, and for the selection of design vehicles. Highway engineers design road geometry to ensure stability of vehicles when negotiating curves and grades and to provide adequate sight distances for undertaking passing maneuvers along curves on two-lane, two-way roads.

Geometric Design

Highway and transportation engineers must meet many safety, service, and performance standards when designing highways for certain site topography. Highway geometric design primarily refers to the visible elements of the highways. Highway engineers who design the geometry of highways must also consider environmental and social effects of the design on the surrounding infrastructure.

There are certain considerations that must be properly addressed in the design process to successfully fit a highway to a site's topography and maintain its safety. Some of these design considerations include:

- Design speed
- Design traffic volume
- Number of lanes
- Level of Service (LOS)
- Sight Distance
- Alignment, super-elevation, and grades
- Cross section
- Lane width
- Horizontal and vertical clearance

The operational performance of a highway can be seen through drivers' reactions to the design considerations and their interaction.

Materials

The materials used for roadway construction have progressed with time, dating back to the early days of the Roman Empire. Advancements in methods with which these materials are characterized and applied to pavement structural design has accompanied this advancement in materials.

There are two major types of pavement surfaces - Portland cement concrete (PCC) and hot-mix asphalt (HMA). Underneath this wearing course are material layers that give structural support for the pavement system. These underlying surfaces may include either the aggregate base and subbase layers, or treated base and subbase layers, and additionally the underlying natural or treated subgrade. These treated layers may be cement-treated, asphalt-treated, or lime-treated for additional support.

Flexible Pavement Design

A flexible, or asphalt, or Tarmac pavement typically consists of three or four layers. For a four layer flexible pavement, there is a surface course, base course, and subbase course constructed over a compacted, natural soil subgrade. When building a three layer flexible pavement, the subbase layer is not used and the base course is placed directly on the natural subgrade.

A flexible pavement's surface layer is constructed of hot-mix asphalt (HMA).Unstabilized aggregates are typically used for the base course; however, the base course could also be stabilized with asphalt, Foamed Bitumen,<Roadstone Recycling> Portland cement, or another stabilizing agent. The subbase is generally constructed from local aggregate material, while the top of the subgrade is often stabilized with cement or lime.

With flexible pavement, the highest stress occurs at the surface and the stress decreases as the depth of the pavement increases. Therefore, the highest quality material needs to be used for the surface, while lower quality materials can be used as the depth of the pavement increases. The term "flexible" is used because of the asphalts ability to bend and deform slightly, then return to its original position as each traffic load is applied and removed. It is possible for these small deformations to become permanent, which can lead to rutting in the wheel path over an extended time.

The service life of a flexible pavement is typically designed in the range of 20 to 30 years. Required thicknesses of each layer of a flexible pavement vary widely depending on the materials used, magnitude, number of repetitions of traffic loads, environmental conditions, and the desired service life of the pavement. Factors such as these are taken into consideration during the design process so that the pavement will last for the designed life without excessive distresses.

Rigid Pavement Design

Rigid pavements are generally used in constructing airports and major highways, such as those in the interstate highway system. In addition, they commonly serve as heavy-duty industrial floor slabs, port and harbor yard pavements, and heavy-vehicle park or terminal pavements. Like flexible pavements, rigid highway pavements are designed as all-weather, long-lasting structures to serve modern day high-speed traffic. Offering high quality riding surfaces for safe vehicular travel, they function as structural layers to distribute vehicular wheel loads in such a manner that the induced stresses transmitted to the subgrade soil are of acceptable magnitudes.

Portland cement concrete (PCC) is the most common material used in the construction of rigid pavement slabs. The reason for its popularity is due to its availability and the economy. Rigid pavements must be designed to endure frequently repeated traffic loadings. The typical designed service life of a rigid pavement is between 30 and 40 years, lasting about twice as long as a flexible pavement.

One major design consideration of rigid pavements is reducing fatigue failure due to the repeated stresses of traffic. Fatigue failure is common among major roads because a typical highway will experience millions of wheel passes throughout its service life. In addition to design criteria such as traffic loadings, tensile stresses due to thermal energy must also be taken into consideration. As pavement design has progressed, many highway engineers have noted that thermally induced stresses in rigid pavements can be just as intense as those imposed by wheel loadings. Due to the relatively low tensile strength of concrete, thermal stresses are extremely important to the design considerations of rigid pavements.

Rigid pavements are generally constructed in three layers - a prepared subgrade, base or subbase, and a concrete slab. The concrete slab is constructed according to a designed choice of plan dimensions for the slab panels, directly influencing the intensity of thermal stresses occurring within the pavement. In addition to the slab panels, temperature reinforcements must be designed to control cracking behavior in the slab. Joint spacing is determined by the slab panel dimensions.

Three main types of concrete pavements commonly used are Jointed plain concrete pavement (JPCP), jointed reinforced concrete pavement (JRCP), and continuously reinforced concrete pavements (CRCP). JPCP's are constructed with contraction joints which direct the natural cracking of the pavement. These pavements do not use any reinforcing steel. JRCP's are constructed with both contraction joints and reinforcing steel to control the cracking of the pavement. High temperatures and moisture stresses within the pavement creates cracking, which the reinforcing steel holds tightly together. At transverse joints, dowel bars are typically placed to assist with transferring the load of the vehicle across the cracking. CRCP's solely rely on continuous reinforcing steel to hold the pavement's natural transverse cracks together. Prestressed concrete pavements have also been used in the construction of highways; however, they are not as common as the other three. Prestressed pavements allow for a thinner slab thickness by partly or wholly neutralizing thermally induced stresses or loadings.

Flexible Pavement Overlay Design

Over the service life of a flexible pavement, accumulated traffic loads may cause excessive rutting or cracking, inadequate ride quality, or an inadequate skid resistance. These problems can be avoided by adequately maintaining the pavement, but the solution usually has excessive maintenance costs, or the pavement may have an inadequate structural capacity for the projected traffic loads.

Throughout a highway's life, its level of serviceability is closely monitored and maintained. One common method used to maintain a highway's level of serviceability is to place an overlay on the pavement's surface.

There are three general types of overlay used on flexible pavements: asphalt-concrete overlay, Portland cement concrete overlay, and ultra-thin Portland cement concrete overlay. The concrete layer in a conventional PCC overlay is placed unbonded on top of the flexible surface. The typical thickness of an ultra-thin PCC overlay is 4 inches (10 cm) or less.

There are three main categories of flexible pavement overlay design procedures:

1. Component Analysis Design
2. Deflection-Based Design

Rigid Pavement Overlay Design

Near the end of a rigid pavement's service life, a decision must be made to either fully reconstruct the worn pavement, or construct an overlay layer. Considering an overlay can be constructed on a rigid pavement that has not reached the end of its service life, it is often more economically attractive to apply overlay layers more frequently. The required overlay thickness for a structurally sound rigid pavement is much smaller than for one that has reached the end of its service life. Rigid and flexible overlays are both used for rehabilitation of rigid pavements such as JPCP, JRCP, and CRCP.

There are three subcategories of rigid pavement overlays that are organized depending on the bonding condition at the pavement overlay and existing slab interface.

1. Bonded overlays
2. Unbonded overlays
3. Partially bonded overlays

Drainage System Design

Designing for proper drainage of highway systems is crucial to their success. Regardless of how well other aspects of a road are designed and constructed, adequate drainage is mandatory for a road to survive its entire service life. Excess water in the highway structure can inevitably lead to premature failure, even if the failure is not catastrophic.

Each highway drainage system is site-specific and can be very complex. Depending on the geography of the region, many methods for proper drainage may not be applicable. The highway engineer must determine which situations a particular design process should be applied, usually a combination of several appropriate methods and materials to direct water away from the structure.

Erosion control is a crucial component in the design of highway drainage systems. Surface drainage must be allowed for precipitation to drain away from the structure. Highways must be designed with a slope or crown so that runoff water will be directed to the shoulder of the road, into a ditch, and away from the site. Designing a drainage system requires the prediction of runoff and infiltration, open channel analysis, and culvert design for directing surface water to an appropriate location.

Construction, Maintenance, and Management

Highway Construction

Highway construction is generally preceded by detailed surveys and subgrade preparation. The methods and technology for constructing highways has evolved over time and become increasingly sophisticated. This advancement in technology has raised the level of skill sets required to manage highway construction projects. This skill varies from project to project, depending on factors such as the project's complexity and nature, the contrasts between new construction and reconstruction, and differences between urban region and rural region projects.

There are a number of elements of highway construction which can be broken up into technical and commercial elements of the system. Some examples of each are listed below:

- Technical Elements
    - Materials
    - Material quality
    - Installation techniques
    - Traffic
- Commercial Elements
    - Contract understanding
    - Environmental aspects
    - Political aspects
    - Legal aspects
    - Public concerns

Typically, construction begins at the lowest elevation of the site, regardless of the project type, and moves upward. By reviewing the geotechnical specifications of the project, information is given about:

- Existing ground conditions
- Required equipment for excavation, grading, and material transportation to and from the site
- Properties of materials to be excavated
- Dewatering requirements necessary for below-grade work
- Shoring requirements for excavation protection
- Water quantities for compaction and dust control

Subbase Course Construction

A subbase course is a layer designed of carefully selected materials that is located between the subgrade and base course of the pavement. The subbase thickness is generally in the range of 4 to 16 inches, and it is designed to withstand the required structural capacity of the pavement section.

Common materials used for a highway subbase include gravel, crushed stone, or subgrade soil that is stabilized with cement, fly ash, or lime. Permeable subbase courses are becoming more prevalent because of their ability to drain infiltrating water from the surface. They also prevent subsurface water from reaching the pavement surface.

When local material costs are excessively expensive or the material requirements to increase the structural bearing of the sub-base are not readily available, highway engineers can increase the bearing capacity of the underlying soil by mixing in Portland cement, foamed asphalt, or use polymer soil stabilization such as cross-linking styrene acrylic polymer that increases the California Bearing Ratio of in-situ materials by a factor 4 – 6.

Base Course Construction

The base course is the region of the pavement section that is located directly under the surface course. If there is a subbase course, the base course is constructed directly about this layer. Otherwise, it is built directly on top of the subgrade. Typical base course thickness ranges from 4 to 6 inches and is governed by underlying layer properties.

Heavy loads are continuously applied to pavement surfaces, and the base layer absorbs the majority of these stresses. Generally, the base course is constructed with an untreated crushed aggregate such as crushed stone, slag, or gravel. The base course material will have stability under the construction traffic and good drainage characteristics.

The base course materials are often treated with cement, bitumen, calcium chloride, sodium chloride, fly ash, or lime. These treatments provide improved support for heavy loads, frost susceptibility, and serves as a moisture barrier between the base and surface layers.

Surface Course Construction

There are two most commonly used types of pavement surfaces used in highway construction: hot-mix asphalt and Portland cement concrete. These pavement surface courses provide a smooth and safe riding surface, while simultaneously transferring the heavy traffic loads through the various base courses and into the underlying subgrade soils.

Hot-mix Asphalt (HMA) Layers

Hot-mix asphalt surface courses are referred to as flexible pavements. The Superpave System was developed in the late 1980s and has offered changes to the design approach, mix design, specifications, and quality testing of materials.

The construction of an effective, long-lasting asphalt pavement requires an experienced construction crew, committed to their work quality and equipment control.

Construction issues:

- Asphalt mix segregation
- Laydown
- Compaction
- Joints

A prime coat is a low viscosity asphalt that is applied to the base course prior to laying the HMA surface course. This coat bonds loose material, creating a cohesive layer between the base course and asphalt surface.

A tack coat is a low viscosity asphalt emulsion that is used to create a bond between an existing pavement surface and new asphalt overlay. Tack coats are typically applied on adjacent pavements (curbs) to assist the bonding of the HMA and concrete.

Portland Cement Concrete (PCC)

Portland cement concrete surface courses are referred to as rigid pavements, or concrete pavements. There are three general classifications of concrete pavements - jointed plain, jointed reinforced, and continuously reinforced.

Traffic loadings are transferred between sections when larger aggregates in the PCC mix inter-lock together, or through load transfer devices in the transverse joints of the surface. Dowel bars are used as load-transferring devices to efficiently transfer loads across transverse joints while maintaining the joint's horizontal and vertical alignment. Tie-bars are deformed steel bars that are placed along longitudinal joints to hold adjacent pavement sections in place.

Highway Maintenance

The overall purpose of highway maintenance is to fix defects and preserve the pavement's structure and serviceability. Defects must be defined, understood, and recorded in order to create an appropriate maintenance plan. Maintenance planning is solving an optimisation problem and it can be predictive. In predictive maintenance planning empirical, data-driven methods give more accurate results than mechanical models. Defects differ between flexible and rigid pavements.

There are four main objectives of highway maintenance:

- repair of functional pavement defects
- extend the functional and structural service life of the pavement
- maintain road safety and signage
- keep road reserve in acceptable condition

Through routine maintenance practices, highway systems and all of their components can be maintained to their original, as-built condition.

Project Management

Project management involves the organization and structuring of project activities from inception to completion. Activities could be the construction of infrastructure such as highways and bridges or major and minor maintenance activities related to constructing such infrastructure. The entire project and involved activities must be handled in a professional manner and completed within deadlines and budget. In addition, minimizing social and environmental impacts is essential to successful project management.

Bridge

The first bridges were made by nature — as simple as a log fallen across a stream. The first bridges made by humans were probably spans of wooden logs or planks and eventually stones, using a simple support and crossbeam arrangement. Most of these early bridges could not support heavy weights or withstand strong currents. It was these inadequacies which led to the development of better bridges. The oldest surviving stone bridge in China is the Zhaozhou Bridge, built from 595 to 605 AD during the Sui Dynasty. This bridge is also historically significant as it is the world's oldest open-spandrel stone segmental arch bridge. The first book on bridge engineering was written by Hubert Gautier in 1716. With the Industrial Revolution in the 19th century, truss systems of wrought iron were developed for larger bridges, but iron did not have the tensile strength to support large loads. With the advent of steel, which has a high tensile strength, much larger bridges were built, many using the ideas of Gustave Eiffel.

Bridges are classified on the basis that how the four forces namely shear, compression, tension, and moment are distributed in the bridge structure.

A structure built to span and provide passage over a river, chasm, road, or any other physical hurdle. The function required from the bridge and the area where it is constructed decides the design of the bridge.

All major bridges are built with the public's money. Therefore, bridge design that best serves the public interest has a threefold goal: to be as efficient, as economical, and as elegant as is safely possible. Efficiency is a scientific principle that puts a value on reducing materials while increasing performance. Economy is a social principle that puts value on reducing the costs of construction and maintenance while retaining efficiency. Finally, elegance is a symbolic or visual principle that puts value on the personal expression of the designer without compromising performance or economy. There is little disagreement over what constitutes efficiency and economy, but the definition of elegance has always been controversial.

Modern designers have written about elegance or aesthetics since the early 19th century, beginning with the Scottish engineer Thomas Telford. Bridges ultimately belong to the general public, which is the final arbiter of this issue, but in general there are three positions taken by professionals. The first principle holds that the structure of a bridge is the province of the engineer and that beauty is fully achieved only by the addition of architecture. The second idea, arguing from the standpoint of pure engineering, insists that bridges making the most efficient possible use of materials are by definition beautiful. The third case holds that architecture is not needed but that engineers must think about how to make the structure beautiful. This last principle recognizes the fact that engineers have many possible choices of roughly equal efficiency and economy and can therefore express their own aesthetic ideas without adding significantly to materials or cost.

Materials

The four primary materials used for bridges have been wood, stone, iron, and concrete. Of these, iron has had the greatest effect on modern bridges. From iron, steel is made, and steel is used to make reinforced and prestressed concrete. Modern bridges are almost exclusively built with steel, reinforced concrete, and prestressed concrete.

Wood and Stone

Wood is relatively weak in both compression and tension, but it has almost always been widely available and inexpensive. Wood has been used effectively for small bridges that carry light loads, such as footbridges. Engineers now incorporate laminated wooden beams and arches into some modern bridges.

Stone is strong in compression but weak in tension. Its primary application has been in arches, piers, and abutments.

Iron and Steel

The first iron used during the Industrial Revolution was cast iron, which is strong in compression but weak in tension. Wrought iron, on the other hand, is as strong in compression as cast iron, but it also has much greater tensile strength. Steel is an even further refinement of iron and is yet stronger, superior to any iron in both tension and compression. Steel can be made to varying strengths, some alloys being five times stronger than others. The engineer refers to these as high-strength steels.

Bridge Maintenance

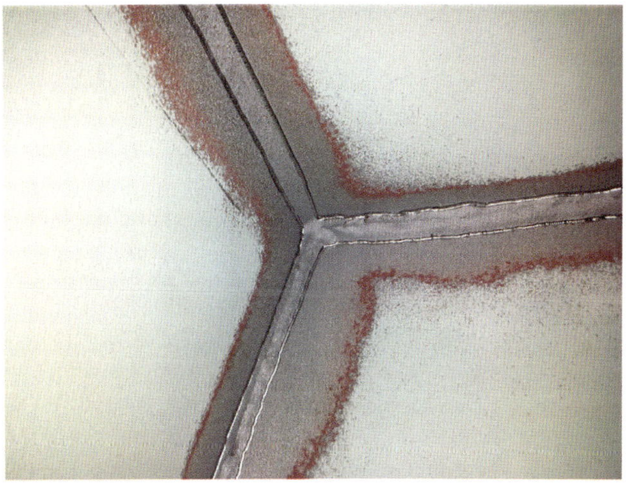

Highway bridge treated with high-frequency impact treatment

Bridge maintenance consisting of a combination of structural health monitoring and testing. This is regulated in country-specific engineer standards and includes e.g. an ongoing monitoring every three to six months, a simple test or inspection every two to three years and a major inspection every six to ten years. In Europe, the cost of maintenance is higher than spending on new bridges. The lifetime of welded steel bridges can be significantly extended by aftertreatment of the weld transitions . This results in a potential high benefit, using existing bridges far beyond the planned lifetime.

Bridge Monitoring

There are several methods used to monitor the stress on large structures like bridges. The most common method is the use of an accelerometer, which is integrated into the bridge while it is being built. This technology is used for long-term surveillance of the bridge.

Another option for structural-integrity monitoring is "non-contact monitoring", which uses the Doppler effect (Doppler shift). A laser beam from a Laser Doppler Vibrometer is directed at the point of interest, and the vibration amplitude and frequency are extracted from the Doppler shift of the laser beam frequency due to the motion of the surface. The advantage of this method is that the setup time for the equipment is faster and, unlike an accelerometer, this makes measurements possible on multiple structures in as short a time as possible. Additionally, this method can measure specific points on a bridge that might be difficult to access.

Tunnel

Tunnels are underground constructions used for transportations. The features, advantages, disadvantage and methods of tunneling in construction is discussed.

The tunnel engineering is one of the most interesting disciplines in engineering. The work is complex and difficult throughout its course, even though it is interesting.

The tunnels are defined as the underground passages that are used for the transportation purposes. These permit the transmission of passengers and freights, or it may be for the transportation of utilities like water, sewage or gas etc.

The operations and the constructions are carried out underground without disturbing the ground surface. This operation is called as the tunneling.

Construction of First Tunnel

It is found that the first tunnel was made by the Egyptians and the Babylonians, 4000 years ago. This tunnel served the purpose of connecting two buildings in Babylon. The connection was from the royal palace to the temple. The length of this tunnel was found to be 910m, which was brick lined.

Selection of Tunneling Route

The two main factors that help in the efficient route of the tunnel are the alignment restraints and the environmental considerations. The underground, as we know is heterogeneous in nature. A proper inspection on the nature of soil, rock, the water table level, and all the alignment restraints had to be made before fixing the route.

The site chosen for tunneling is such a way that the inconvenience and difficulty that is caused to the environment in that area including living is minimum.

The tunneling method chosen depends on the ground conditions, the water table level, the tunnel drive length and the diameter, the tunnel depth, final utility requirements, the shape of the tunnel and the risk of construction.

Advantages of Tunneling

The tunneling method gain certain advantages compared with other methods, which are mentioned below:

- The tunneling procedure is more economical in nature, compared to open cut trench method when the depth is beyond a limit

- The surface life or ground activities like transportation are not disturbed when tunneling is undergone.
- The method ensures high-speed construction with low power consumption
- Reduces Noise Pollution
- These methods have freedom from snow and iceberg hazards, in areas of high altitudes
- Surface and air interference is restricted for tunnels
- Provision of tunnels with easy gradients, help in reducing the cost of hauling
- For the transportation of public utilities, tunneling method has a remarkable advantage compared to the bridge.
- The dangerous open cut to a nearby structure, when it is needed, is solved by the tunneling method
- The tunneling grant greater protection in aerial warfare and bombing conditions

Tunneling Disadvantages

The tunneling method gains certain disadvantages, which is due to its complexity and difficulty. Some of them are:

- The initial investment cost for commencing the tunnel is high compared to the open cut method.
- Highly skilled and experienced designers and engineer team only will work best for this operation.
- Higher and constant supervision from the start to the end of the tunneling project is necessary without any compromise
- Highly sophisticated and specialized equipment are necessary to perform the tunneling operations.

Construction

Tunnels are dug in types of materials varying from soft clay to hard rock. The method of tunnel construction depends on such factors as the ground conditions, the ground water conditions, the length and diameter of the tunnel drive, the depth of the tunnel, the logistics of supporting the tunnel excavation, the final use and shape of the tunnel and appropriate risk management.

There are three basic types of tunnel construction in common use. Cut-and-cover tunnels are constructed in a shallow trench and then covered over. Bored tunnels are constructed in situ, without removing the ground above. Finally a tube can be sunk into a body of water, which is called an immersed tunnel.

Cut-and-cover

Cut-and-cover is a simple method of construction for shallow tunnels where a trench is excavated and roofed over with an overhead support system strong enough to carry the load of what is to be built above the tunnel. Two basic forms of cut-and-cover tunnelling are available:

- *Bottom-up method*: A trench is excavated, with ground support as necessary, and the tunnel is constructed in it. The tunnel may be of in situ concrete, precast concrete, precast arches, or corrugated steel arches; in early days brickwork was used. The trench is then carefully back-filled and the surface is reinstated.
- *Top-down method*: Side support walls and capping beams are constructed from ground level by such methods as slurry walling or contiguous bored piling. Then a shallow excavation allows making the tunnel roof of precast beams or in situ concrete. The surface is then reinstated except for access openings. This allows early reinstatement of roadways, services and other surface features. Excavation then takes place under the permanent tunnel roof, and the base slab is constructed.

Cut-and-cover construction of the Paris Métro in France

Shallow tunnels are often of the cut-and-cover type (if under water, of the immersed-tube type), while deep tunnels are excavated, often using a tunnelling shield. For intermediate levels, both methods are possible.

Large cut-and-cover boxes are often used for underground metro stations, such as Canary Wharf tube station in London. This construction form generally has two levels, which allows economical arrangements for ticket hall, station platforms, passenger access and emergency egress, ventilation and smoke control, staff rooms, and equipment rooms. The interior of Canary Wharf station has been likened to an underground cathedral, owing to the sheer size of the excavation. This contrasts with many traditional stations on London Underground, where bored tunnels were used for stations and passenger access. Nevertheless, the original parts of the London Underground network, the Metropolitan and District Railways, were constructed using cut-and-cover. These lines pre-dated electric traction and the proximity to the surface was useful to ventilate the inevitable smoke and steam.

A major disadvantage of cut-and-cover is the widespread disruption generated at the surface level during construction. This, and the availability of electric traction, brought about London Underground's switch to bored tunnels at a deeper level towards the end of the 19th century.

Boring Machines

Tunnel boring machines (TBMs) and associated back-up systems are used to highly automate the entire tunnelling process, reducing tunnelling costs. In certain predominantly urban applications, tunnel boring is viewed as quick and cost effective alternative to laying surface rails and roads. Expensive compulsory purchase of buildings and land, with potentially lengthy planning inquiries, is eliminated. Disadvantages of TBMs arise from their usually large size – the difficulty of transporting the large TBM to the site of tunnel construction, or (alternatively) the high cost of assembling the TBM on-site, often within the confines of the tunnel being constructed.

A workman is dwarfed by the tunnel boring machine used to excavate the Gotthard Base Tunnel (Switzerland), the world's longest railway tunnel.

There are a variety of TBM designs that can operate in a variety of conditions, from hard rock to soft water-bearing ground. Some types of TBMs, the bentonite slurry and earth-pressure balance machines, have pressurised compartments at the front end, allowing them to be used in difficult conditions below the water table. This pressurizes the ground ahead of the TBM cutter head to balance the water pressure. The operators work in normal air pressure behind the pressurised compartment, but may occasionally have to enter that compartment to renew or repair the cutters. This requires special precautions, such as local ground treatment or halting the TBM at a position free from water. Despite these difficulties, TBMs are now preferred over the older method of tunnelling in compressed air, with an air lock/decompression chamber some way back from the TBM, which required operators to work in high pressure and go through decompression procedures at the end of their shifts, much like deep-sea divers.

In February 2010, Aker Wirth delivered a TBM to Switzerland, for the expansion of the Linth–Limmern Power Stations located south of Linthal in the canton of Glarus. The borehole has a diameter of 8.03 metres (26.3 ft). The four TBMs used for excavating the 57-kilometre (35 mi) Gotthard Base Tunnel, in Switzerland, had a diameter of about 9 metres (30 ft). A larger TBM was built to bore the Green Heart Tunnel (Dutch: Tunnel Groene Hart) as part of the HSL-Zuid in the Netherlands, with a diameter of 14.87 metres (48.8 ft). This in turn was superseded by the Madrid M30 ringroad, Spain, and the Chong Ming tunnels in Shanghai, China. All of these machines were built at least partly by Herrenknecht. As of August 2013, the world's largest TBM is "Big Bertha", a 57.5-foot (17.5 m) diameter machine built by Hitachi Zosen Corporation, which is digging the Alaskan Way Viaduct replacement tunnel in Seattle, Washington (US).

Clay-kicking

Clay-kicking is a specialised method developed in the United Kingdom of digging tunnels in strong clay-based soil structures. Unlike previous manual methods of using mattocks which relied on the soil structure to be hard, clay-kicking was relatively silent and hence did not harm soft clay-based structures. The clay-kicker lies on a plank at a 45-degree angle away from the working face and inserts a tool with a cup-like rounded end with the feet. Turning the tool manually, the kicker extracts a section of soil, which is then placed on the waste extract.

Used in Victorian civil engineering, the method found favour in the renewal of Britain's ancient sewerage systems, by not having to remove all property or infrastructure to create a small tunnel system. During the First World War, the system was used by Royal Engineer tunnelling companies to put mines beneath the German Empire lines. The method was virtually silent and so not susceptible to listening methods of detection.

Shafts

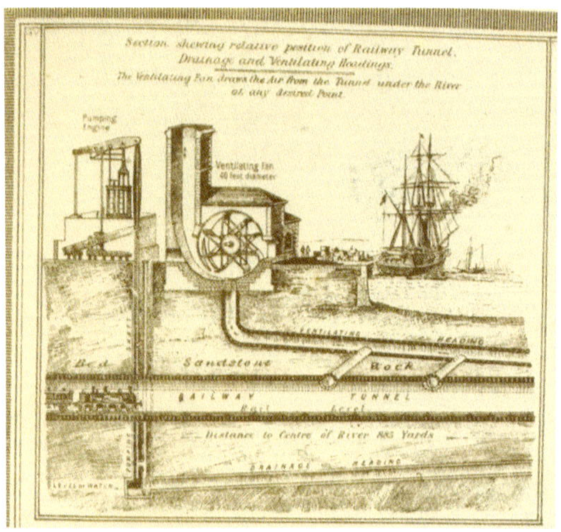

illustration showing the ventilation and drainage system of the Mersey railway tunnel

A temporary access shaft is sometimes necessary during the excavation of a tunnel. They are usually circular and go straight down until they reach the level at which the tunnel is going to be built. A shaft normally has concrete walls and is usually built to be permanent. Once the access shafts are complete, TBMs are lowered to the bottom and excavation can start. Shafts are the main entrance in and out of the tunnel until the project is completed. If a tunnel is going to be long, multiple shafts at various locations may be bored so that entrance to the tunnel is closer to the unexcavated area.

Once construction is complete, construction access shafts are often used as ventilation shafts, and may also be used as emergency exits.

Sprayed Concrete Techniques

The New Austrian Tunneling Method (NATM) was developed in the 1960s and is the best known of a number of engineering practices that use calculated and empirical measurements to provide safe support to the tunnel lining. The main idea of this method is to use the geological stress of the surrounding rock mass to stabilize the tunnel, by allowing a measured relaxation and stress reassignment into the surrounding rock to prevent full loads becoming imposed on the supports. Based on geotechnical measurements, an optimal cross section is computed. The excavation is protected by a layer of sprayed concrete, commonly referred to as shotcrete. Other support measures can include steel arches, rockbolts and mesh. Technological developments in sprayed concrete technology have resulted in steel and polypropylene fibres being added to the concrete mix to improve lining strength. This creates a natural load-bearing ring, which minimizes the rock's deformation.

Illowra Battery utility tunnel, Port Kembla. One of many bunkers south of Sydney.

By special monitoring the NATM method is flexible, even at surprising changes of the geomechanical rock consistency during the tunneling work. The measured rock properties lead to appropriate tools for tunnel strengthening. In the last decades also soft ground excavations up to 10 kilometres (6.2 mi) became usual.

Pipe Jacking

In pipe jacking, hydraulic jacks are used to push specially made pipes through the ground behind a TBM or shield. This method is commonly used to create tunnels under existing structures, such as roads or railways. Tunnels constructed by pipe jacking are normally small diameter bores with a maximum size of around 3.2 metres (10 ft).

Box Jacking

Box jacking is similar to pipe jacking, but instead of jacking tubes, a box-shaped tunnel is used. Jacked boxes can be a much larger span than a pipe jack, with the span of some box jacks in excess of 20 metres (66 ft). A cutting head is normally used at the front of the box being jacked, and spoil removal is normally by excavator from within the box.Recent developments of the Jacked Arch and Jacked deck have enabled longer and larger structures to be installed to close accuracy. The 126m long 20m clear span underpass below the high speed rail lines at Cliffsend in Kent, UK is an example of this technique.

Underwater Tunnels

There are also several approaches to underwater tunnels, the two most common being bored tunnels or immersed tubes, examples are Bjørvika Tunnel and Marmaray. Submerged floating tunnels are a novel approach under consideration; however, no such tunnels have been constructed to date.

Shark tunnel at the Georgia Aquarium

Temporary Way

During construction of a tunnel it is often convenient to install a temporary railway, particularly to remove excavated spoil, often narrow gauge so that it can be double track to allow the operation of empty and loaded trains at the same time. The temporary way is replaced by the permanent way at completion, thus explaining the term "Perway".

Enlargement

A utility tunnel in Prague

The vehicles or traffic using a tunnel can outgrow it, requiring replacement or enlargement:

- The original single line Gib Tunnel near Mittagong was replaced with a double-track tunnel, with the original tunnel used for growing mushrooms.
- The 1832 double-track mile-long tunnel from Edge Hill to Lime Street in Liverpool was near totally removed, apart from a 50-metre section at Edge Hill and a section nearer to Lime Street, as four tracks were required. The tunnel was dug out into a very deep four-track cutting, with short tunnels in places along the cutting. Train services were not interrupted as the work progressed. There are other occurrences of tunnels being replaced by open cuts, for example, the Auburn Tunnel.

- The Farnworth Tunnel in England was enlarged using a tunnel boring machine (TBM) in 2015. The Rhyndaston Tunnel was enlarged using a borrowed TBM so as to be able to take ISO containers.
- Tunnels can also be enlarged by lowering the floor.

Open Building Pit

An open building pit consists of a horizontal and a vertical boundary that keeps groundwater and soil out of the pit. There are several potential alternatives and combinations for (horizontal and vertical) building pit boundaries. The most important difference with cut-and-cover is that the open building pit is muted after tunnel construction; no roof is placed.

Other Construction Methods

- Drilling and blasting
- Hydraulic splitter
- Slurry-shield machine
- Wall-cover construction method.

Roads

Road Engineering

Since the beginning of the 20th century, as the automobile and truck have offered ever higher levels of mobility, vehicle ownership per head of population has increased. Road needs have been strongly influenced by this popularity and also by the mass movement of people to cities and thence to suburban fringes—a trend that has led to increasing travel needs and road congestion and to low-density cities, which are difficult to service by public transport. Often the building of new roads to alleviate such problems has encouraged further urban sprawland yet more road travel. Long-term solutions require the provision of alternatives to car and truck transport, controls over land use, and the proper pricing of road travel. To this end, road managers must be concerned not merely with lines on maps but also with the number, type, speed, and loading of individual vehicles, the safety, comfort, and convenience of the traveling public, and the health and welfare of bystanders and adjoining property owners.

Ideally, the development of a major road system is an orderly, continuous process. The process follows several steps: assessing road needs and transport options; planning a system to meet those needs; designing an economically, socially, and environmentally acceptable set of roads; obtaining the required approval and financing; building, operating, and maintaining the system; and providing for future extensions and reconstruction.

Planning

Road needs are closely associated with the relative location of centres of population, commerce, industry, and transportation. Trafficbetween two centres is approximately proportional to their populations and inversely proportional to the distance between them. Estimating traffic on a route thus requires a prediction of future population growth and economic activity, an estimation of their effects on land use and travel needs, and a knowledge of any potential transport alternatives. The key variables defining road needs are the traffic volumes, tonnages, and speeds to be expected throughout the road's life.

Once the traffic demand has been estimated, it is necessary to predict the extent of the road works needed to handle that traffic. A starting point in these calculations is offered by surveys of the origins, destinations, and route choices of present traffic; computer models are then used to estimate future traffic volumes on each proposed route. Estimates of route choice are based on the understanding that most drivers select their estimate of the quickest, shortest, or cheapest route. Consideration in planning is also given to the effect of new traffic on existing streets, roads, and parking provisions.

Where feasible, the next step in planning a road system is to refine the selected route to a narrow corridor. The various alignment options are drawn, considering the local terrain and conditions. The economic, social, and environmental benefits and costs of these options are discussed with relevant official and community groups until an acceptable specific route is determined.

Road Design

Alignment and Profile

After a route has been selected, a three-dimensional road alignment and its associated cross-sectional profiles are produced. In order to reduce the amount of earth to be moved, the alignment is adjusted where practical so that the earth to be excavated is in balance with the embankments to be built. Computers allow many options to be explored and realistic views of the future road to be examined.

In order to fully understand the design stage, a few standard terms must be defined. A traffic lane is the portion of pavement allocated to a single line of vehicles; it is indicated on the pavement by painted longitudinal lines or embedded markers. The shoulder is a strip of pavement outside an outer lane; it is provided for emergency use by traffic and to protect the pavement edges from traffic damage. A set of adjoining lanes and shoulders is called a roadway or carriageway, while the pavement, shoulders, and bordering roadside up to adjacent property lines are known as the right-of-way.

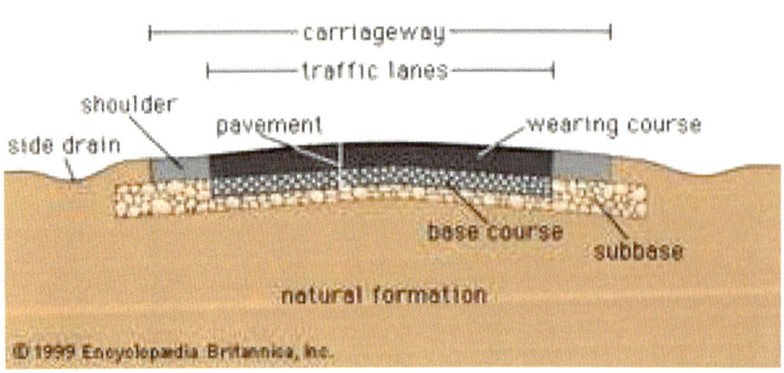

Schematic cross section of a modern roadway.*Encyclopædia Britannica, Inc.*

In order to maintain quality and uniformity, design standards are established for each functional road type. The number of traffic lanes is directly determined by the combination of traffic volume and speed, since practical limits on vehicle spacing means that there is a maximum number of vehicles per hour that pass through a traffic lane. The width of lanes and shoulders, which must strike a balance between construction cost and driver comfort, allows the carriageway width to be determined. Standards also specify roadside barriers or give the clear transverse distances needed on either side of the carriageway in order to provide safety in the event that vehicles accidentally leave the carriageway. Thus it is possible to define the total right-of-way width needed for the entire road, although intersections will add further special demands.

Design standards also help to determine the actual alignment of the road by specifying, for each design speed, the minimum radius of horizontal curves, the maximum vertical gradient, the clearance under bridges, and the distance a driver must be able to see the pavement ahead in order to stop or turn aside.

Construction

Surveyor at work with a leveling instrument

Asphalt layer and Hamm road roller

"Roadbed" (from the 'related terms' Bed (geology) & streambed), as well as the term "Road building"

Road constructionrequires the creation of an engineered continuous right-of-way or roadbed, overcoming geographic obstacles and having grades low enough to permit vehicle or foot travel, and may be required to meet standards set by law or official guidelines. The process is often begun with the removal of earth and rock by digging or blasting, construction of embankments, bridges and tunnels, and removal of vegetation (this may involve deforestation) and followed by the laying of pavement material. A variety of road building equipment is employed in road building.

After design, approval, planning, legal and environmentalconsiderations have been addressed alignment of the road is set out by a surveyor. The radii and gradient are designed and staked out to best suit the natural ground levels and minimize the amount of cut and fill. Great care is taken to preserve reference Benchmarks

Sub-base layer composed of cement-based material being applied during construction of the M8 motorway in Ireland

Roads are designed and built for primary use by vehicular and pedestrian traffic. Storm drainage and environmental considerations are a major concern. Erosion and sediment controls are constructed to prevent detrimental effects. Drainage lines are laid with sealed joints in the road easement with runoff coefficientsand characteristics adequate for the land zoning and storm water system. Drainage systems must be capable of carrying the ultimate design flow from the upstream catchment with approval for the outfall from the appropriate authority to a watercourse, creek, river or the sea for drainage discharge.

A borrow pit (source for obtaining fill, gravel, and rock) and a water source should be located near or in reasonable distance to the road construction site. Approval from local authorities may be required to draw water or for working (crushing and screening) of materials for construction needs. The topsoil and vegetation is removed from the borrow pit and stockpiled for subsequent rehabilitation of the extraction area. Side slopes in the excavation area not steeper than one vertical to two horizontal for safety reasons.

Old road surfaces, fences, and buildings may need to be removed before construction can begin. Trees in the road construction area may be marked for retention. These protected trees should not have the topsoil within the area of the tree's drip line removed and the area should be kept clear of construction material and equipment. Compensation or replacement may be required if a protected tree is damaged. Much of the vegetation may be mulched and put aside for use during reinstatement. The topsoil is usually stripped and stockpiled nearby for rehabilitation of newly constructed embankments along the road. Stumps and roots are removed and holes filled as required before the earthwork begins. Final rehabilitation after road construction is completed will include seeding, planting, watering and other activities to reinstate the area to be consistent with the untouched surrounding areas.

Processes during earthwork include excavation, removal of material to spoil, filling, compacting, construction and trimming. If rock or other unsuitable material is discovered it is removed, moisture content is managed and replaced with standard fill compacted to meet the design requirements (generally 90-95% relative compaction). Blasting is not frequently used to excavate the roadbed as the intact rock structure forms an ideal road base. When a depression must be filled to come up to the road grade the native bed is compacted after the topsoil has been removed. The fill is made by the "compacted layer method" where a layer of fill is spread then compacted to specifications, under saturated conditions. The process is repeated until the desired grade is reached.

Typical pavement strata for a heavily traveled road

General fill material should be free of organics, meet minimum California bearing ratio (CBR) results and have a low plasticity index. The lower fill generally comprises sand or a sand-rich mixture with fine gravel, which acts as an inhibitor to the growth of plants or other vegetable matter. The compacted fill also serves as lower-stratum drainage. Select second fill (sieved) should be composed of gravel, decomposed rock or broken rockbelow a specified particle size and be free of large lumps of clay. Sand clay fill may also be used. The roadbed must be "proof rolled" after each layer of fill is compacted. If a roller passes over an area without creating visible deformation or spring the section is deemed to comply.

Geosynthetics such as geotextiles, geogrids and geocells are frequently used in the various pavement layers to improve road quality. These materials and methods are used in low-traffic private roadways as well as public roads and highways. Geosynthetics perform four main functions in roads: separation, reinforcement, filtration and drainage; which increase the pavement performance, reduce construction costs and decrease maintenance.

The completed road way is finished by paving or left with a gravel or other natural surface. The type of road surface is dependent on economic factors and expected usage. Safety improvements such as traffic signs, crash barriers, raised pavement markers and other forms of road surface marking are installed.

According to a May 2009 report by the American Association of State Highway and Transportation Officials (AASHTO) and TRIP – a national transportation research organization – driving on rough roads costs the average American motorist approximately $400 a year in extra vehicle operating costs. Drivers living in urban areas with populations more than 250,000 are paying upwards of $750 more annually because of accelerated vehicle deterioration, increased maintenance, additional fuel consumption, and tire wear caused by poor road conditions.

When a single carriageway road is converted into dual carriageway by building a second separate carriageway alongside the first, it is usually referred to as *duplication*, *twinning* or *doubling*. The original carriageway is changed from two-way to become one-way, while the new carriageway is one-way in the opposite direction. In the same way as converting railway lines from single track to double track, the new carriageway is not always constructed directly alongside the existing carriageway.

Geometric Design of Roads or Highways

Geometric design of highway is the determination of layout and features visible on highway. The emphasis is more on satisfying the need of the driver as well as to ensure the safety of the vehicle, the comfort while driving and efficiency. Other related factors are also considered based on the project.

The main features considered for geometric design of highway are:

- Cross section elements
- The gradient
- The intersection
- The consideration of sight distance

The geometric design of highway is influenced by:

- The characteristics of the vehicle
- The behavior of the driver
- The psychology of the driver Traffic characteristics
- Traffic Volume
- Traffic Speed

Severity of movement and accidents can be reduced largely by implementing a proper design. The main objective of geometric design is to get optimum efficiency in the traffic operation period and maximum safety.

All these features must be attained with maximum economy in the cost and construction. Unlike the construction of pavement, the planning process is carried out in advance.

Factors Affecting the Geometric Design of Highways

The various factors that govern the geometric design of highways are:

- The Design Speed
- The topography factors
- Other Factors

Design Speed of Highways

The most important factor affecting the geometric design of the highway is the design speed. The design speed parameter affects the:

- Length of the vertical curves
- The horizontal curves
- The sight distances

Speed is a parameter that have huge variation will the person driving, the vehicle type, the topography etc. This is the reason why the design speed is taken as a primary factor in the geometric design of highways.

Design Standards

Roads are designed in conjunction with design guidelines and standards. These are adopted by national and sub-national authorities (e.g., states, provinces, territories and municipalities). Design guidelines take into account speed, vehicle type, road grade (slope), view obstructions, and stopping distance. With proper application of guidelines, along with good engineering judgement, an engineer can design a roadway that is comfortable, safe, and appealing to the eye.

The primary US guidance is found in *A Policy on Geometric Design of Highways and Streets* published by the American Association of State Highway and Transportation Officials (AASHTO). Other standards include the Australian Guide to Road Design, and the British Design Manual for Roads. An open source version of the green book is published online by The Council for Scientific and Industrial Research (CSIR) office in Zimbabwe.

Profile

The profile of a road consists of road slopes, called grades, connected by parabolic vertical curves. Vertical curves are used to provide a gradual change from one road slope to another, so that vehicles may smoothly navigate grade changes as they travel.

Sag vertical curves are those that have a tangent slope at the end of the curve that is higher than that of the beginning of the curve. When driving on a road, a sag curve would appear as a valley, with the vehicle first going downhill before reaching the bottom of the curve and continuing uphill or level.

Crest vertical curves are those that have a tangent slope at the end of the curve that is lower than that of the beginning of the curve. When driving on a crest curve, the road appears as a hill, with the vehicle first going uphill before reaching the top of the curve and continuing downhill.

The profile also affects road drainage. Very flat roads and sag curves may have poor drainage, and steep roads have high velocity flows.

Terminology

- BVC = Beginning of Vertical Curve

- EVC = End of Vertical Curve
- g_1 = initial roadway grade, expressed in percent
- g_2 = final roadway grade, expressed in percent
- A = absolute value of the difference in grades (initial minus final), expressed in percent
- h_1 = Height of eye above roadway, measured in meters or feet
- h_2 = Height of object above roadway, measured in meters or feet
- L = curve length (along the x-axis)
- PVI = point of vertical interception (intersection of initial and final grades)
- tangent elevation = elevation of a point along the initial tangent
- x = horizontal distance from BVC
- Y (offset) = vertical distance from the initial tangent to a point on the curve
- Y' = curve elevation = tangent elevation - offset

Sag Curves

Sag vertical curves are curves which, when viewed from the side, are concave upwards. This includes vertical curves at valley bottoms, but it also includes locations where an uphill grade becomes steeper, or a downhill grade becomes less steep.

The most important design criterion for these curves is headlight sight distance. When a driver is driving on a sag curve at night, the sight distance is limited by the higher grade in front of the vehicle. This distance must be long enough that the driver can see any obstruction on the road and stop the vehicle within the headlight sight distance. The headlight sight distance (S) is determined by the angle of the headlight and angle of the tangent slope at the end of the curve. By first finding the headlight sight distance (S) and then solving for the curve length (L) in each of the equations below, the correct curve length can be determined. If the S<L curve length is greater than the headlight sight distance, then this number can be used. If it is smaller, this value cannot be used. Similarly, if the S>L curve length is smaller than the headlight sight distance, then this number can be used. If it is larger, this value cannot be used.

Units	**Sight Distance < Curve Length (S<L)**	**Sight Distance > Curve Length (S>L)**
Metric	$L = \frac{AS^2}{120 + 3.5S}$	$L = 2S - \frac{120 + 3.5S}{A}$
US Customary	$L = \frac{AS^2}{400 + 3.5S}$	$L = 2S - \frac{400 + 3.5S}{A}$

These equations assume that the headlights are 600 millimetres (2.0 ft) above the ground, and the headlight beam diverges 1 degree above the longitudinal axis of the vehicle.

Crest Curves

Crest vertical curves are curves which, when viewed from the side, are convex upwards. This includes vertical curves at hill crests, but it also includes locations where an uphill grade becomes less steep, or a downhill grade becomes steeper.

The most important design criterion for these curves is stopping sight distance. This is the distance a driver can see over the crest of the curve. If the driver cannot see an obstruction in the roadway, such as a stalled vehicle or an animal, the driver may not be able to stop the vehicle in time to avoid a crash. The desired stopping sight distance (S) is determined by the speed of traffic on a road. By first finding the stopping sight distance (S) and then solving for the curve length (L) in each of the equations below, the correct curve length can be determined. The proper equation depends on whether the vertical curve is shorter or longer than the available sight distance. Normally, both equations are solved, then the results are compared to the curve length.

Sight Distance > Curve Length (S>L)

$$L = 2S - \frac{200(\sqrt{h_1} + \sqrt{h_2})^2}{A}$$

Sight Distance < Curve Length (S<L)

$$L = \frac{AS^2}{200(\sqrt{h_1} + \sqrt{h_2})^2}$$

US standards specify the height of the driver's eye is defined as 1080 mm (3.5 ft) above the pavement, and the height of the object the driver needs to see as 600 mm (2.0 ft), which is equivalent to the taillight height of most passenger cars.

For bicycle facilities, the cyclist's eye height is assumed to be at 1.4 m (4.5 ft), and the object height is 0 inches, since a pavement defect can cause a cyclist to fall or lose control.

Alignment

Horizontal alignment in road design consists of straight sections of road, known as tangents, connected by circular horizontal curves. Circular curves are defined by radius (tightness) and deflection angle (extent). The design of a horizontal curve entails the determination of a minimum radius (based on speed limit), curve length, and objects obstructing the view of the driver.

Using AASHTO standards, an engineer works to design a road that is safe and comfortable. If a horizontal curve has a high speed and a small radius, an increased superelevation (bank) is needed in order to assure safety. If there is an object obstructing the view around a corner or curve, the engineer must work to ensure that drivers can see far enough to stop to avoid an accident or accelerate to join traffic.

Terminology

- R = Radius

- PC = Point of Curvature (point at which the curve begins)
- PT = Point of Tangent (point at which the curve ends)
- PI = Point of Intersection (point at which the two tangents intersect)
- T = Tangent Length
- C = Long Chord Length (straight line between PC and PT)
- L = Curve Length
- M = Middle Ordinate, now known as HSO - Horizontal Sightline Offset (distance from sight-obstructing object to the middle of the outside lane)
- E = External Distance
- f_s = Coefficient of Side Friction
- u = Vehicle Speed
- Δ = Deflection Angle

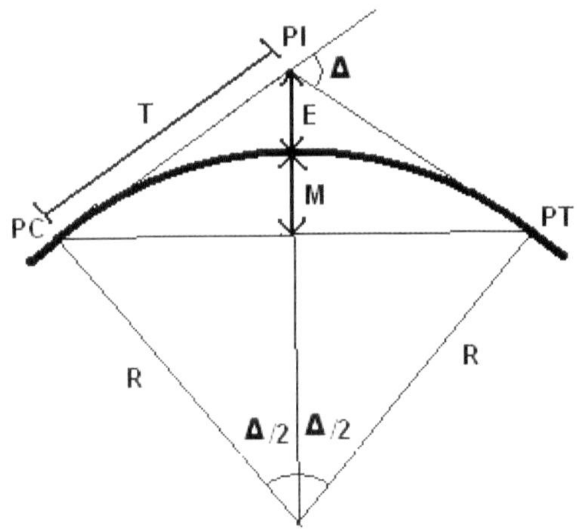

Geometry

$$T = R\tan\left(\frac{\Delta}{2}\right)$$

$$C = 2R\sin\left(\frac{\Delta}{2}\right)$$

$$L = R\pi\frac{\Delta}{180}$$ [2]

$$M = R\left(1-\cos\left(\frac{\Delta}{2}\right)\right)$$

$$E = R\left(\frac{1}{\cos\left(\frac{\Delta}{2}\right)} - 1\right)$$

Curve Sight Distance

$$M = R\left(1 - \cos\left(\frac{28.65S}{R}\right)\right)$$

Cross Section

The cross section of a roadway can be considered a representation of what one would see if an excavator dug a trench across a roadway, showing the number of lanes, their widths and cross slopes, as well as the presence or absence of shoulders, curbs, sidewalks, drains, ditches, and other roadway features. The cross-sectional shape of a road surface, in particular in connection to its role in managing runoff, is called "crown".

Lane Width

The selection of lane width affects the cost and performance of a highway. Typical lane widths range from 3 metres (9.8 ft) to 3.6 metres (12 ft). Wider lanes and shoulders are usually used on roads with higher speed and higher volume traffic, and significant numbers of trucks and other large vehicles. Narrower lanes may be used on roads with lower speed or lower volume traffic.

Narrow lanes cost less to build and maintain, but also reduce the capacity of a road to convey traffic. On rural roads, narrow lanes are likely to experience higher rates of run-off-road and head-on collisions. Wider roads increase the time needed to walk across, and increase stormwater runoff.

Cross Slope

Cross slope describes the slope of a roadway perpendicular to the centerline. If a road were completely level, water would drain off it very slowly. This would create problems with hydroplaning, and ice accumulation in cold weather.

In tangent (straight) sections, the road surface cross slope is commonly 1-2% to enable water to drain from the roadway. Cross slopes of this size, especially when applied in both directions of travel with a crown point along the centerline of a roadway are commonly referred to as "normal crown" and are generally unnoticeable to traveling motorists.

In curved sections, the outside edge of the road is *superelevated* above the centerline. Since the road is sloped down toward the inside of the curve, gravity draws the vehicle toward the inside of the curve. This causes a greater proportion of centripetal force to supplant the tyre friction that would otherwise be needed to negotiate the curve.

Superelevation slopes of 4 to 10% are applied in order to aid motorists in safely traversing these sections, while maintaining vehicle speed throughout the length of the curve. An upper bound of 12% was chosen to meet the demands of construction and maintenance practices, as well as to limit the difficulty of driving a steeply cross-sloped curve at low speeds. In areas that receive significant snow and ice, most agencies use a maximum cross slope of 6 to 8%. While steeper cross slope makes it difficult to traverse the slope at low speed when the surface is icy, and when accelerating from zero with warm tyres on the ice, lower cross slope increases the risk of loss-of-control at high speeds, especially when the surface is icy. Since the consequence of high speed skidding is much worse than that of sliding inward at a low speed, sharp curves have the benefit of greater net safety when designers select up to 8% superelevation, instead of 4%. A lower slope of 4% is commonly used on urban roadways where speeds are lower, and where a steeper slope would raise the outside road edge above adjacent terrain.

The equation for the desired radius of a curve, shown below, takes into account the factors of speed and superelevation (e). This equation can be algebraically rearranged to obtain desired rate of superelevation, using input of the roadway's designated speed and curve radius.

$$R = \frac{u^2}{15(e + f_s)}$$

The American Association of State Highway and Transportation officials (AASHTO) provides a table from which desired superelevation rates can be interpolated, based on the designated speed and radius of a curved section of roadway. This table can also be found in many state roadway design guides and manuals in the U.S.

Recent research has shown that, considering rollover risk for heavy vehicles (semitrailers & buses), which have a relatively high centre-of-gravity, the above equation yields cross slope values which are too low.

Safety Effects of Road Geometry

The geometry of a road influences its safety performance. While studies of contributing factors to road accidents show that human factors predominate, roadway factors are the second most common category, with vehicle factors last.

Design Consistency

Collisions tend to be more frequent in locations where a sudden change in road character violates the driver's expectations. A common example is a sharp curve at the end of a long tangent section of road. The concept of *design consistency* addresses this by comparing adjacent road segments and identifying sites with changes the driver might find sudden or unexpected. Locations with large changes in the predicted operating speed are likely to benefit from additional design effort. A horizontal curve with a significantly smaller radius than those before it may need enhanced curve signs. This is an improvement on the concept of design speed, which only sets a lower limit for geometric design. In the example given above, a long tangent followed by a sharp curve would be acceptable if a 30 mph design speed was chosen. Design consistency analysis would flag the decrease in operating speed at the curve.

Safety Effects of Alignment

The safety of a horizontal curve is affected by the length of the curve, the curve radius, whether spiral transition curves are used, and the superelevation of the roadway. For a given curve deflection, crashes are more likely on curves with a smaller radius. Spiral transitions decrease crashes, and insufficient superelevation increases crashes.

A safety performance function to model curve performance on two-lane roads is:bharat

$$AMF = \frac{1.55L_c + \frac{80.2}{R} - .012S}{1.55L_c}$$

where

AMF = Accident modification factor, a multiplier that describes how many more crashes are likely to occur on the curve compared to a straight road

L_c = Length of the horizontal curve in miles.

R = Radius of the curve in feet.

S = 1 if spiral transition curves are present

= 0 if spiral transition curves are absent

Safety Effects of Cross Section

Cross slope and lane width affect the safety performance of a road.

Certain types of crashes, termed "lane departure crashes", are more likely on roads with narrow lanes. These include run-off-road collisions, sideswipes, and head-on collisions. For two-lane rural roads carrying over 2000 vehicles per day, the expected increase in crashes is:

Lane width	Expected increase in crashes
12 feet (3.7 m)	0%
11 feet (3.4 m)	5%
10 feet (3.0 m)	30%
9 feet (2.7 m)	50%

The effect of lane width is reduced on urban and suburban roads and low volume roads.

Insufficient superelevation will also result in an increase in crash rate. The expected increase is shown below:

Superelevation deficiency	Expected increase in crashes for cars	Expected increase in crashes for heavy trucks
<0.01	0%	<5%
0.02	6%	10%

0.03	9%	15%
0.04	12%	20%
0.05	15%	25%

Sight Distance

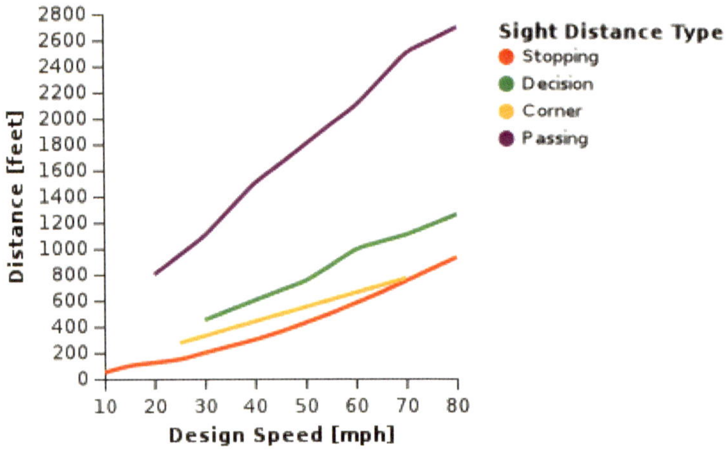

Sight Distance, by Type

Road geometry affects the sight distance available to the driver. Sight distance, in the context of road design, is defined as "the length of roadway ahead visible to the driver." Sight distance is how far a road user (usually a vehicle driver) can see before the line of sight is blocked by a hill crest, or an obstacle on the inside of a horizontal curve or intersection. Insufficient sight distance can adversely affect the safety or operations of a roadway or intersection.

The sight distance needed for a given situation is the distance travelled during the two phases of a driving maneuver: perception-reaction time (PRT), and maneuver time (MT). Perception-reaction time is the time it takes for a road user to realize that a reaction is needed to a road condition, decided what maneuver is appropriate, and start the maneuver. Maneuver time is the time it takes to complete the maneuver. The distance driven during perception-reaction time and maneuver time is the sight distance needed.

During highway design and traffic safety investigations, highway engineers compare the available sight distance to how much sight distance is needed for the situation. Depending on the situation, one of three types of sight distances will be used.

Stopping Sight Distance

Stopping sight distance is the distance traveled during perception-reaction time (while the vehicle driver perceives a situation requiring a stop, realizes that stopping is necessary, and applies the brake), and maneuver time (while the driver decelerates and comes to a stop). Actual stopping distances are also affected by road conditions, the mass of the car, the incline of the road, and numerous other factors. For design, a conservative distance is needed to allow a vehicle traveling at design speed to stop before reaching a stationary object in its path. Typically the design sight distance allows a below-average driver to stop in time to avoid a collision.

Decision Sight Distance

Decision sight distance is used when drivers must make decisions more complex than stop or don't stop. It is longer than stopping sight distance to allow for the distance traveled while making a more complex decision. The decision sight distance is "distance required for a driver to detect an unexpected or otherwise difficult-to-perceive information source or hazard in a roadway environment that may be visually cluttered, recognize the hazard or its threat potential, select an appropriate speed and path, and initiate and complete the required maneuver safely and efficiently". Ideally, roads are designed for the decision sight distance, using 6 to 10 seconds for perception-reaction time and 4 to 5 seconds to perform the right maneuver.

Intersection Sight Distance

Intersection sight distance is the sight distance needed to safely proceed through an intersection. The distance needed depends on the type of traffic control at the intersection (uncontrolled, yield sign, stop sign or signal), and the maneuver (left turn, right turn, or proceeding straight). All-way stop intersections need the least, and uncontrolled intersections require the most. Intersection sight distance is a key factor in whether no control or yield control can be safely used, or more restrictive control in needed.

Corner Sight Distance

Corner sight distance (CSD) is the road alignment specification which provides a substantially clear line of sight so that the driver of a vehicle, bicyclist or pedestrian waiting at the crossroad may safely anticipate the driver of an approaching vehicle. Corner sight provides an adequate time for the waiting user to either cross all lanes of through traffic, cross the near lanes and turn left, or turn right, without requiring through traffic to radically alter their speed.

Uncontrolled and Yield Controlled Intersections

Uncontrolled and yield (give way) controlled intersections require large sight triangles clear of obstructions in order to operate safely. At uncontrolled intersections, the basic right-of-way rules apply (either yield to the vehicle on the right, or the boulevard rule, depending on the location). Vehicle drivers must be able to see traffic approaching on the intersecting road at a point where they can adjust their speed, or stop if need be, to yield to the other traffic before reaching the intersection. It isn't the only criterion for allowing these types of intersection control. Changing an intersection to stop control is a common response to poor safety performance.

Two-way Stop Control

When determining corner sight distance, a set back distance for the vehicle waiting at the crossroad must be assumed. Set back for the driver of the vehicle on the crossroad has been standardized by some state MUTCDs and design manuals to be up to a minimum of 10 feet plus the shoulder width of the major road but not less than 15 feet. However, the Federal MUTCD requires that a stop line, if used, shall be at least 4 feet from the nearest travel lane. Line of sight for corner sight distance is to be determined from a 3 and 1/2-foot eye height at the vehicle driver's location on the minor road to a 4 and 1/4-foot object height in the center of the approaching lane of the major road. Corner sight distance, D_{CSD}, is equivalent to a specified time gap, t_g, at the design speed, V_{DS}, required for a stopped vehicle to turn right or left:

$$D_{CSD} = V_{DS} t_g$$

For passenger vehicles at two lane intersections, this time gap equivalence is commonly a distance 7.5 seconds away at the design speed. Longer gaps are required for trucks and buses, and for multilane roads. Generally, the public right-of-way should include and maintain this line-of-sight.

All-way Stop Control and Signalized Intersections

Drivers at intersections with all-way stop control or traffic signals need the least sight distance. At all-way stops, drivers need to be able to see vehicles stopped at other approaches. At signals, drivers approaching the intersections need to see the signal heads. In jurisdictions that allow right turn on red, drivers in the right lane stop control need the same sight distance as two-way stop control. Although not needed during normal operations, additional sight distance should be provided for signal malfunctions and power outages.

Effects of Insufficient Sight Distance

Many roads were created long before the current sight distance standards were adopted, and the financial burden on many jurisdictions would be formidable to: acquire and maintain additional right-of-way; redesign roadbeds on all of them; or implement future projects on rough terrain, or environmentally sensitive areas. In such cases, the bare minimum corner sight distance should be equal to the stopping sight distance. While a corner sight distance which far exceed the braking distance at the design speed should be afforded to the driver, he or she is still generally required to maintain such control and safe speed as to be able to stop within the Assured Clear Distance Ahead (ACDA), and the basic speed rule always applies. Jurisdictions often provide some level of design immunity against government claims actions, in such cases.

Warning signs are often used where sight distance is insufficient. The Manual on Uniform Traffic Control Devices requires Stop Ahead, Yield Ahead or Signal Ahead signs at intersections where the traffic control device is not visible from a distance equal to the stopping sight distance at speed of approaching traffic. Hill Blocks View signs can be used where crest vertical curves restrict sight distance. However, many jurisdictions still expect drivers to use ordinary care regarding conditions readily apparent to a driver, without the prompting of a sign. The care and focus ordinarily required of a driver against certain types of hazards may be somewhat amplified on roads with lower functional classification. The probability of spontaneous traffic increases proportionally to the density of access points, and this density should be readily apparent to a driver even when a specific access point is not. For this reason, full corner sight distance is almost never required for individual driveways in urban high-density residential areas, and street parking is commonly permitted within the right-of-way.

Stopping Sight Distance

The safe and efficient operation of vehicles on the road depends very much on the visibility of the road ahead of the driver. Thus the geometric design of the road should be done such that any obstruction on the road length could be visible to the driver from some distance ahead . This distance is said to be the sight distance.

Stopping sight distance (SSD) is the minimum sight distance available on a highway at any spot having sufficient length to enable the driver to stop a vehicle traveling at design speed, safely without collision with any other obstruction.

There is a term called *safe stopping distance* and is one of the important measures in traffic engineering. It is the distance a vehicle travels from the point at which a situation is first perceived to the time the deceleration is complete. Drivers must have adequate time if they are to suddenly respond to a situation. Thus in highway design, sight distance atleast equal to the safe stopping distance should be provided. The stopping sight distance is the sum of lag distance and the braking distance. Lag distance is the distance the vehicle traveled during the reaction time t and is given by vt, where v is the velocity in m/sec^2.

Braking distance is the distance traveled by the vehicle during braking operation. For a level road this is obtained by equating the work done in stopping the vehicle and the kinetic energy of the vehicle. If F is the maximum frictional force developed and the braking distance is l, then work done against friction in stopping the vehicle is $Fl = fWl$. where W is the total weight of the vehicle. The kinetic energy at the design speed is

$$\frac{1}{2}mv^2 = \frac{1}{2}\frac{Wv^2}{g}$$

$$fWl = \frac{Wv^2}{2g}$$

$$l = \frac{v^2}{2gf}$$

Therefore, the SSD = lag distance + braking distance and given by:

$$SSD = vt + \frac{v^2}{2gf}$$

where v is the design speed in m/sec^2, t is the reaction time in $sec, g is$ is the acceleration due to gravity and f is the coefficient of friction. The coefficient of friction f is given below for various design speed.

Table: Coefficient of longitudinal friction					
Speed, kmph	< 30	40	50	60	> 80
f	0.40	0.38	0.37	0.36	0.35

When there is an ascending gradient of say $+n\%$, the component of gravity adds to braking action and hence braking distance is decreased. The component of gravity acting parallel to the surface which adds to the the braking force is equal to $W \sin\alpha \approx W \tan\alpha = Wn/100$. Equating kinetic energy and work done:

$$\left(fW + \frac{Wn}{100}\right)l = \frac{Wv^2}{2g}$$

$$l = \frac{v^2}{2g\left(f + \frac{n}{100}\right)}$$

Similarly the braking distance can be derived for a descending gradient. Therefore the general equation is given by Equation 2.

$$SSD = vt + \frac{v^2}{2g(f \pm 0.01n)}$$

1. At night. If the object is not illuminated and there is no street lighting then you can only see as far as your headlights. If the object is another vehicle and its rear lights are on, you will see beyond your headlights but may take longer to perceive what's going on and to react to it.

2. If the vehicle is not a car. Typical heavy vehicle stopping distances for heavy vehicles can be up to 2.75 times those of a car (Lay,ref. 1635), whilst stopping distances for bicyclists is of course much less. Most guidelines quote SSD for cars.This problem can be reduced if heavy vehicles are subject to "vehicle type" speed limits.

3. On curves. Here some of the available friction is taken up by radial friction which helps prevent the vehicle spinning off the road.

4 In three-dimensional space. Where a combination of longitudinal gradient, superelevation and vehicle length is not allowed for in the basic formulae.

Where the application of the theory varies

- Different values for the input parameters such as perception-reaction times, deceleration, friction
- Different values for perception-reaction time (Lay Ref. 1635 page 428 refers to 2.5 secs in rural areas and 1.5 secs for urban areas)
- Assumption of different values for friction with different vehicle speeds
- Different types of vehicle
- Different definitions of speed V

Typical Values

Table: Some values for the input parameters

Parameter		Value	Ref.
perception time (secs)	Tp	1,5	167
reaction time (secs)	Tr	1	167
deceleration (m/sec^2)	--	3,4	1639
friction	f	0,38	

Table 2: Some values for SSD (against speed in km/hr)

Country	Ref.	Year	50	60	70	80	100	120
Germany (motorways, level gradient)	1615	2008	54	71	90	111	160	217
USA	1006	1994	57	74	94	112	157	202

Health Warning

The values for SSD quoted in guidelines are usually for cars driving on wet roads with an asphalt or concrete surface, in daylight and good visibility, with a comfortable level of braking deceleration, and with various secondary provisos (car brakes and tyres in good condition, 90th %ile driver (ref. 1639 "a deceleration rate of 3.4 m/sec2 is considered to be comfortable for 90% of drivers")and not impaired by e.g. alcohol,and so on.)

Design Speed

Horizontal alignment is one of the most important features influencing the efficiency and safety of a highway. A poor design will result in lower speeds and resultant reduction in highway performance in terms of safety and comfort. In addition, it may increase the cost of vehicle operations and lower the highway capacity. Horizontal alignment design involves the understanding on the design aspects such as design speed and the effect of horizontal curve on the vehicles. The horizontal curve design elements include design of super elevation, extra widening at horizontal curves, design of transition curve, and set back distance.

The design speed, as noted earlier, is the single most important factor in the design of horizontal alignment. The design speed also depends on the type of the road. For e.g, the design speed expected from a National highway will be much higher than a village road, and hence the curve geometry will vary significantly.

The design speed also depends on the type of terrain. A plain terrain can afford to have any geometry, but for the same standard in a hilly terrain requires substantial cutting and filling implying exorbitant costs as well as safety concern due to unstable slopes. Therefore, the design speed is normally reduced for terrains with steep slopes.

For instance, Indian Road Congress (IRC) has classified the terrains into four categories, namely plain, rolling, mountainous, and steep based on the cross slope as given in table below. Based on the type of road and type of terrain the design speed varies. The IRC has suggested desirable or ruling speed as well as minimum suggested design speed and is tabulated in table.

Table: Terrain classification	
Terrain classification	Cross slope (%)
Plain	0-10
Rolling	10-25
Mountainous	25-60
Steep	>60

The recommended design speed is given in Table.

Table: Design speed in *km/hr* as per IRC (ruling and minimum)				
Type	Plain	Rolling	Hilly	Steep
NS&SH	100-80	80-65	50-40	40-30
MDR	80-65	65-50	40-30	30-20
ODR	65-50	50-40	30-25	25-20
VR	50-40	40-35	25-20	25-20

Geometric Features

The design speed chosen for a highway is a major factor in choosing superelevation rates and radii of curves, sight distance, and the lengths of crest and sag vertical curves. Roads with higher travel speeds require sweeping curves, steeper curve banking, longer sight distances, and more gentle hill crests and valleys. Lower speed roads can have sharper curves, less banking, less sight distance, and sharper hill crests and valleys.

Speed Limit Variance from Design Speed

While a road's design speed is sometimes used to determine an initial speed limit, it is an imperfect measure of the maximum speed at which a motor vehicle can be operated for reasons including:

- It is only a theoretical or laboratory measurement created before a road is even built.
- The highest design speed for a road or segment is the design speed of its least favorable part. For example, given a road segment with a 60 mph design speed except for a curve with a 45 mph design speed, the entire segment would have a 45 mph design speed. In reality, the road may have a 45 mph advisory speed on the curve and higher safe operating speeds elsewhere.
- The design speed may be higher than legislated speed limit caps, so it would not be legal to sign some roads at their design speeds.
- It is based on the capabilities of vehicles and roadways that existed at or before the design speed was determined. Vehicular and roadway technologies generally improve over time. Therefore, as time elapses from when a roadway's original design speed was determined, it is increasingly likely that a design speed will underestimate the maximum safe speed.

Recognizing the limitations on the use of the design speed for speed limit determination, "operating speeds and even posted speed limits can be higher than design speeds without necessarily compromising safety".

Evolution

The concept of design speed is evolving. The definition in the 1994 edition of the AASHTO *Green Book,* was "the maximum safe speed that can be maintained over a specified section of highway when conditions are so favorable that the design features of the highway govern. The assumed design speed should be a logical one with respect to the topography, the adjacent land use, and the functional classification of highway." A majority of US states used this definition. In 2004, the first sentence was changed to "a selected speed used to determine the various geometric design features of a roadway." This reflects the fact that meeting a minimum design speed is not enough to ensure a safe roadway.

Recently, the concept of design consistency has been used instead of minimum design speeds. This attempts to connect driver's expectations about the roadway with the roadway design. It uses driver behavior models to predict vehicle speeds on highway segments, and compares the predicted speed on adjacent segments. Significant reductions in speed from one segment to the next are flagged as locations where drivers may end up driving too fast for road conditions.

A major shift in philosophy is also taking place regarding design speed of urban and suburban streets. Highway engineers would measure the prevailing speed on a road, round up to the next multiple of 5 mph, and design the road for that speed, assuming that it would be safe. Recent research and design practices have focused on using the street design to influence drivers to choose a speed appropriate for the neighborhood. This speed has been called the "target speed," and is ideally equal to the speed limit.

Factors

When roads are planned, the selected design speed may be based on or influence several factors, including:

- geometric design of road features
- planned operating speed
- legislated speed limit caps
- anticipated traffic volume
- the road's functional classification

Road Surface

A paved road surface, commonly constructed of bituminous material, should provide a durable, even running surface with adequate skid resistance which is able to withstand the effects of weather.

When the road surface condition becomes poor or defective, rehabilitation or resurfacing work is needed to restore the pavement surface to the required level of service.

Not all roads are the same. Although the driver rarely gets to set foot on the road surface as they move by in their vehicle there is actually a huge variety of surfaces that their tyres make contact with.

Underneath the vehicle, there are many different types of road surfacing material, each chosen for a particular site depending on countless variables. The components of the road surface can be influenced by:

- the type of road
- the volume and speed of traffic
- the location amongst others.

Getting these variables right and applying the correct surface option can literally be a matter of life and death.

Asphalt

Closeup of asphalt on a driveway

Asphalt (specifically, asphalt concrete), sometimes called flexible pavement due to the nature in which it distributes loads, has been widely used since the 1920s. The viscous nature of the bitumen binder allows asphalt concrete to sustain significant plastic deformation, although fatigue from repeated loading over time is the most common failure mechanism. Most asphalt surfaces are laid on a gravel base, which is generally at least as thick as the asphalt layer, although some 'full depth' asphalt surfaces are laid directly on the native subgrade. In areas with very soft or expansive subgrades such as clay or peat, thick gravel bases or stabilization of the subgrade with Portland cement or lime may be required. Polypropylene and polyester geosynthetics have also been used for this purpose and in some northern countries, a layer of polystyrene boards have been used to delay and minimize frost penetration into the subgrade.

Depending on the temperature at which it is applied, asphalt is categorized as hot mix, warm mix, or cold mix. Hot mix asphalt is applied at temperatures over 300 °F (150 °C) with a free floating screed. Warm mix asphalt is applied at temperatures of 200–250 °F (95–120 °C), resulting in reduced energy usage and emissions of volatile organic compounds. Cold mix asphalt is often used on lower-volume rural roads, where hot mix asphalt would cool too much on the long trip from the asphalt plant to the construction site.

An asphalt concrete surface will generally be constructed for high-volume primary highways having an average annual daily traffic load greater than 1200 vehicles per day. Advantages of asphalt roadways include relatively low noise, relatively low cost compared with other paving methods, and perceived ease of repair. Disadvantages include less durability than other paving methods, less tensile strength than concrete, the tendency to become slick and soft in hot weather and a certain amount of hydrocarbon pollution to soil and groundwater or waterways.

In the mid-1960s, rubberized asphalt was used for the first time, mixing crumb rubber from used tires with asphalt. While a potential use for tires that would otherwise fill landfills and present a fire hazard, rubberized asphalt has shown greater incidence of wear in freeze-thaw cycles in temperate zones due to non-homogeneous expansion and contraction with non-rubber components. The application of rubberized asphalt is more temperature-sensitive, and in many locations can only be applied at certain times of the year.

Study results of the long-term acoustic benefits of rubberized asphalt are inconclusive. Initial application of rubberized asphalt may provide 3–5 decibels (dB) reduction in tire-pavement source noise emissions; however, this translates to only 1–3 decibels (dB) in total traffic noise level reduction (due to the other components of traffic noise). Compared to traditional passive attenuating measures (e.g., noise walls and earth berms), rubberized asphalt provides shorter-lasting and lesser acoustic benefits at typically much greater expense.

Concrete

Concrete roadway in San Jose, California

Concrete surfaces (specifically, Portland cement concrete) are created using a concrete mix of Portland cement, coarse aggregate, sand and water. In virtually all modern mixes there will also be various admixtures added to increase workability, reduce the required amount of water, mitigate harmful chemical reactions and for other beneficial purposes. In many cases there will also be Portland cement substitutes added, such as fly ash. This can reduce the cost of the concrete and improve its physical properties. The material is applied in a freshly mixed slurry, and worked mechanically to compact the interior and force some of the cement slurry to the surface to produce a smoother, denser surface free from honeycombing. The water allows the mix to combine molecularly in a chemical reaction called hydration.

A concrete road in Ewing, New Jersey. The original pavement was laid in the 1950s and has not been significantly altered since.

Concrete surfaces have been refined into three common types: jointed plain (JPCP), jointed reinforced (JRCP) and continuously reinforced (CRCP). The one item that distinguishes each type is the jointing system used to control crack development.

A concrete road in Ewing, New Jersey. The original pavement was laid and has not been significantly altered since

One of the major advantages of concrete pavements is they are typically stronger and more durable than asphalt roadways. They also can be grooved to provide a durable skid-resistant surface. A notable disadvantage is that they typically can have a higher initial cost, and can be more time-consuming to construct. This cost can typically be offset through the long life cycle of the pavement. Concrete pavement can be maintained over time utilizing a series of methods known as concrete pavement restoration which include diamond grinding, dowel bar retrofits, joint and crack sealing, cross-stitching, etc. Diamond grinding is also useful in reducing noise and restoring skid resistance in older concrete pavement.

The first street in the United States to be paved with concrete was Court Avenue in Bellefontaine, Ohio in 1893. The first mile of concrete pavement in the United States was on Woodward Avenue in Detroit, Michigan in 1909. Following these pioneering uses, the Lincoln Highway Association, established in October 1913 to oversee the creation of one of the United States' earliest east-west transcontinental highways for the then-new automobile, began to establish "seedling miles" of specifically concrete-paved roadbed in various places in the American Midwest, starting in 1914 west of Malta, Illinois, while using concrete with the specified concrete "ideal section" for the Lincoln Highway in Lake County, Indiana during 1922 and 1923.

Composite Pavement

An example of composite pavement: hot-mix asphalt overlaid onto Portland cement concrete pavement

Composite pavements combine a Portland cement concrete sublayer with an asphalt. They are usually used to rehabilitate existing roadways rather than in new construction.

Asphalt overlays are sometimes laid over distressed concrete to restore a smooth wearing surface. A disadvantage of this method is that movement in the joints between the underlying concrete slabs, whether from thermal expansion and contraction, or from deflection of the concrete slabs from truck axle loads, usually causes *reflective cracks* in the asphalt. To decrease reflective cracking, concrete pavement is broken apart through a *break and seat, crack and seat*, or *rubblization* process. Geosynthetics can be used for reflective crack control. With break and seat and crack and seat processes, a heavy weight is dropped on the concrete to induce cracking, then a heavy roller is used to seat the resultant pieces into the subbase. The main difference between the two processes is the equipment used to break the concrete pavement and the size of the resulting pieces. The theory is frequent small cracks will spread thermal stress over a wider area than infrequent large joints, reducing the stress on the overlying asphalt pavement. Rubblization is a more complete fracturing of the old, worn-out concrete, effectively converting the old pavement into an aggregate base for a new asphalt road.

Whitetopping uses Portland cement concrete to resurface a distressed asphalt road.

Recycling

Distressed road materials can be reused when rehabilitating a roadway. The existing pavement is ground or broken up into small pieces, through a process called milling. It can then be transported to an asphalt or concrete plant and incorporated into new pavement, or recycled in place to form the base or subbase for new pavement. Some methods used include:

In-place Recycling

- Rubblizing of concrete pavement. Existing concrete pavement is broken into gravel-sized particles. Any steel reinforcing is removed, then the remaining gravel-sized particles are compacted and overlaid with asphalt pavement.
- Cold in-place recycling. Bituminous pavement is ground or milled into small particles. The asphalt millings are blended with a small amount of asphalt emulsion or foamed bitumen, paved and compacted, allowed to cure for seven to ten days, then overlaid with asphalt.

- Hot in-place recycling. Bituminous pavement is heated to 250 to 300 °F (120 to 150 °C), milled, combined with a rejuvenating agent or virgin asphalt binder, and compacted. It may then be overlaid with a new asphalt overlay. This process only recycles the top two inches (50 mm) or less, so it can be used to correct rutting, polishing or other surface defects. It is not a good procedure for roads with structural failures. It also generates high heat and vapor emissions, and may not be a good candidate for built-up areas.
- Full depth reclamation is a process which pulverizes the full thickness of the asphalt pavement and some of the underlying material to provide a uniform blend of material. A binding agent may be mixed in to form a base course for the new pavement, or it may be left unbound to form a subbase course. Common binding agents include asphalt emulsion, fly ash, Portland cement or calcium chloride. It can also be mixed with aggregate, recycled asphalt millings, or crushed Portland cement to improve the gradation of the material, and can provide a design life cycle of 30 years with proper lab testing and field verification.

Bituminous Surface

Bituminous surface treatment (BST) or chipseal is used mainly on low-traffic roads, but also as a sealing coat to rejuvenate an asphalt concrete pavement. It generally consists of aggregate spread over a sprayed-on asphalt emulsion or cut-back asphalt cement. The aggregate is then embedded into the asphalt by rolling it, typically with a rubber-tired roller. This type of surface is described by a wide variety of regional terms including "chip seal", "tar and chip", "oil and stone", "seal coat", "sprayed seal" or "surface dressing" or as simply "bitumen."

BST is used on hundreds of miles of the Alaska Highway and other similar roadways in Alaska, the Yukon Territory, and northern British Columbia. The ease of application of BST is one reason for its popularity, but another is its flexibility, which is important when roadways are laid down over unstable terrain that thaws and softens in the spring.

Other types of BSTs include micropaving, slurry seals and Novachip. These are laid down using specialized and proprietary equipment. They are most often used in urban areas where the roughness and loose stone associated with chip seals is considered undesirable.

Thin Membrane Surface

A thin membrane surface (TMS) is an oil-treated aggregate which is laid down upon a gravel road bed, producing a dust-free road. A TMS road reduces mud problems and provides stone-free roads for local residents where loaded truck traffic is negligible. The TMS layer adds no significant structural strength, and so is used on secondary highways with low traffic volume and minimal weight loading. Construction involves minimal subgrade preparation, following by covering with a 50-to-100-millimetre (2.0–3.9 in) cold mix asphalt aggregate. The Operation Division of the Ministry of Highways and Infrastructure in Saskatchewan has the responsibility of maintaining 6,102 kilometres (3,792 mi) of thin membrane surface (TMS) highways.

Otta Seal

Otta seal is a low-cost road surface using a 16–30-millimetre (0.63–1.18 in) thick mixture of bitumen and crushed rock.

Gravel Surface

Gravel is known to have been used extensively in the construction of roads by soldiers of the Roman Empire but in 1998 a limestone-surfaced road, thought to date back to the Bronze Age, was found at Yarnton in Oxfordshire, Britain. Applying gravel, or "metalling," has had two distinct usages in road surfacing. The term *road metal* refers to the broken stone or cinders used in the construction or repair of roads or railways, and is derived from the Latin *metallum*, which means both "mine" and "quarry". The term originally referred to the process of creating a gravel roadway. The route of the roadway would first be dug down several feet and, depending on local conditions, French drains may or may not have been added. Next, large stones were placed and compacted, followed by successive layers of smaller stones, until the road surface was composed of small stones compacted into a hard, durable surface. "Road metal" later became the name of stone chippings mixed with tar to form the road surfacing material tarmac. A road of such material is called a "metalled road" in Britain, a "paved road" in Canada and the US, or a "sealed road" in parts of Canada, Australia and New Zealand.

A granular surface can be used with a traffic volume where the annual average daily traffic is 1,200 vehicles per day or less. There is some structural strength if the road surface combines a sub base and base and is topped with a double graded seal aggregate with emulsion. Besides the 4,929 kilometres (3,063 mi) of granular pavements maintained in Saskatchewan, around 40% of New Zealand roads are unbound granular pavement structures.

The decision whether to pave a gravel road or not often hinges on traffic volume. It has been found that maintenance costs for gravel roads often exceed the maintenance costs for paved or surface-treated roads when traffic volumes exceed 200 vehicles per day.

Some communities are finding it makes sense to convert their low-volume paved roads to aggregate surfaces.

Other Surfaces

Pavers (or paviours), generally in the form of pre-cast concrete blocks, are often used for aesthetic purposes, or sometimes at port facilities that see long-duration pavement loading. Pavers are rarely used in areas that see high-speed vehicle traffic.

Brick, cobblestone, sett, wood plank, and wood block pavements such as Nicolson pavement, were once common in urban areas throughout the world, but fell out of fashion in most countries, due to the high cost of labor required to lay and maintain them, and are typically only kept for historical or aesthetic reasons. In some countries, however, they are still common in local streets. In the Netherlands, brick paving has made something of a comeback since the adoption of a major nationwide traffic safety program in 1997. From 1998 through 2007, more than 41,000 km of city streets were converted to local access roads with a speed limit of 30 km/h, for the purpose of traffic calming. One popular measure is to use brick paving - the noise and vibration slows motorists down. At the same time, it is not uncommon for cycle paths alongside a road to have a smoother surface than the road itself.

Likewise, macadam and tarmac pavements can still sometimes be found buried underneath asphalt concrete or Portland cement concrete pavements, but are rarely constructed today.

Concrete pavers

There are also other methods and materials to create pavements that have appearance of brick pavements. The first method to create brick texture is to heat an asphalt pavement and use metal wires to imprint a brick pattern using a compactor to create stamped asphalt. A similar method is to use rubber imprinting tools to press over a thin layer of cement to create decorative concrete. Another method is to use a brick pattern stencil and apply a surfacing material over the stencil. Materials that can be applied to give the color of the brick and skid resistance can be in many forms. An example is to use colored polymer-modified concrete slurry which can be applied by screeding or spraying. Another material is aggregate-reinforced thermoplastic which can be heat applied to the top layer of the brick-pattern surface. Other coating materials over stamped asphalt are paints and two-part epoxy coating.

Replacing the old road with concrete blocks in Bo'ao Road area, Haikou City, Hainan, China

Polymer cement overlaying to change asphalt pavement to brick texture and color to create decorative crosswalk

Acoustical implications

Cobbles

Roadway surfacing choices are known to affect the intensity and spectrum of sound emanating from the tire/surface interaction. Initial applications of noise studies occurred in the early 1970s. Noise phenomena are highly influenced by vehicle speed.

Roadway surface types contribute differential noise effects of up to 4 dB, with chip seal type and grooved roads being the loudest, and concrete surfaces without spacers being the quietest. Asphaltic surfaces perform intermediately relative to concrete and chip seal. Rubberized asphalt has been shown to give a marginal 3–5 dB reduction in tire-pavement noise emissions, and a marginally discernible 1–3 dB reduction in total road noise emissions when compared to conventional asphalt applications.

Rectangles

Decorative wavy pattern on La Rambla.

Decorative mock-brick pattern.

More decorative brickwork patterns.

Surface Deterioration

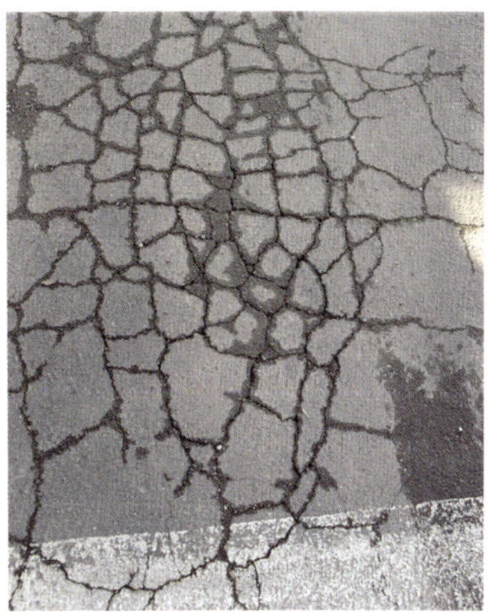

Deteriorating asphalt

As pavement systems primarily fail due to fatigue (in a manner similar to metals), the damage done to pavement increases with the fourth power of the axle load of the vehicles traveling on it. According to the AASHO Road Test, heavily loaded trucks can do more than 10,000 times the damage done by a normal passenger car. Tax rates for trucks are higher than those for cars in most countries for this reason, though they are not levied in proportion to the damage done. Passenger cars are considered to have little practical effect on a pavement's service life, from a materials fatigue perspective.

Other failure modes include aging and surface abrasion. As years go by, the binder in a bituminous wearing course gets stiffer and less flexible. When it gets "old" enough, the surface will start losing aggregates, and macrotexture depth increases dramatically. If no maintenance action is done quickly on the wearing course, potholes will form. The freeze-thaw cycle in cold climates will dramatically accelerate pavement deterioration, once water can penetrate the surface.

If the road is still structurally sound, a bituminous surface treatment, such as a chipseal or surface dressing can prolong the life of the road at low cost. In areas with cold climate, studded tires may be allowed on passenger cars. In Sweden and Finland, studded passenger car tires account for a very large share of pavement rutting.

The physical properties of a stretch of pavement can be tested using a falling weight deflectometer.

Several design methods have been developed to determine the thickness and composition of road surfaces required to carry predicted traffic loads for a given period of time. Pavement design methods are continuously evolving. Among these are the Shell Pavement design method, and the American Association of State Highway and Transportation Officials (AASHTO) 1993 "Guide for Design of Pavement Structures". A new mechanistic-empirical design guide has been under development by NCHRP (Called Superpave Technology) since 1998. A new design guide called Mechanistic Empirical Pavement Design Guide (MEPDG) was developed and is about to be adopted by AASHTO.

Further research by University College London into pavements has led to the development of an indoor, 80-sq-metre artificial pavement at a research centre called Pedestrian Accessibility and Movement Environment Laboratory (PAMELA). It is used to simulate everyday scenarios, from different pavement users to varying pavement conditions. There also exists a research facility near Auburn University, the NCAT Pavement Test Track, that is used to test experimental asphalt pavements for durability.

In addition to repair costs, the condition of a road surface has economic effects for road users. Rolling resistance increases on rough pavement, as does wear and tear of vehicle components. It has been estimated that poor road surfaces cost the average US driver $324 per year in vehicle repairs, or a total of $67 billion. Also, it has been estimated that small improvements in road surface conditions can decrease fuel consumption between 1.8 and 4.7%.

Markings

Road surface markings are used on paved roadways to provide guidance and information to drivers and pedestrians. It can be in the form of mechanical markers such as cat's eyes, botts' dots and rumble strips, or non-mechanical markers such as paints, thermoplastic, plastic and epoxy.

Drainage Gradient

Most paving[1] and drainage relies on fall to carry away the water to somewhere less troublesome. Fall may also be referred to as slope or, more correctly, gradient. By making one part of the pavement or drainage pipe higher than another, gravity can be relied upon to persuade the water in a preferred direction.

But how much fall is enough? And can you have too much?

Types of Fall:

Endfall and crossfall

The term 'fall' is often used in relation to paving and drainage, but it is sometimes further qualified by talking about endfall and crossfall.

Endfall:

This is also known more corectly as longitudinal fall and it describes the fall or gradient along the length of a pavement or surface.

Crossfall:

Known more correctly as transverse fall, it describes the fall or gradient across the breadth of a pavement or surface.

On some pavements, notably those covering larger areas such as plazas and bigger car parks, it's not always possible to determine just which direction is endfall or which is crossfall. It may be that the project engineer has arbitrarily decided that one particular orientation is longitudinal and so is represented as endfall, and perpendicular to that will be the crossfall.

Understanding Gradient

There needs to be some universally understood method for describing how much slope or fall there is on any given structure, such as a pavement or a pipe. The way this is achieved is by measuring how much that structure rises or falls over a specified distance.

It should be apparent that a pipe measuring 1m in length which is 100mm higher at one end than the other (blue) has more fall than a pipe with 50mm difference over the same distance (red), or a pipe that has that 100mm of difference over a 1.5m length (yellow).

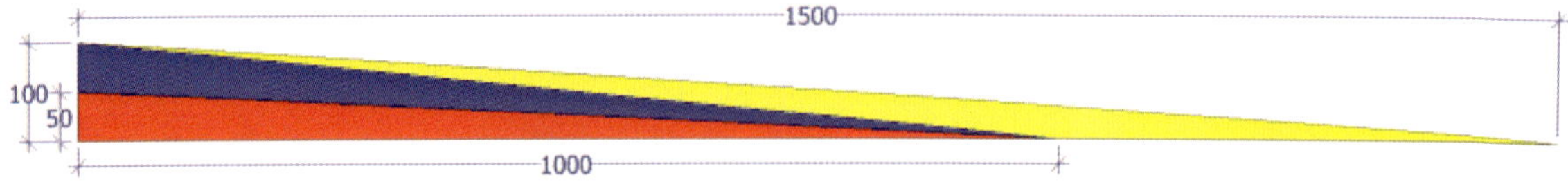

These two values, namely amount of difference in level and the distance over which that difference occurs, are used to define any given gradient. In the useful easy-to-remember equation used for the remainder of this page, the level difference is referred to as "Rise" (or fall) while the distance is referred to as "Run".

$$Gradient = Rise \div Run$$

$$gradient = \frac{Rise}{Run}$$

Minimum Drainage Gradient

Most road design manuals require drainage gradient to exceed 0.5%, in order to drain water and prevent excessive skid accidents.

One exception to the minimum 0.5% DG limit can be found in the Norwegian road design manual, where the minimum drainage gradient is 2% instead of 0.5%.

Typical Values of Drainage Gradient

Typically on straight road sections, the drainage gradient is at least 1–3% due to the normal cross slope of 1–3%.

In curved sections the drainage gradient is higher, and may often reach 5–12% due to superelevated CS that may reach 5–8% in areas with icy roads and up to 12% in areas without icy roads.

The road's longitudinal slope contributes to a high drainage gradient. However, longitudinal highway slopes steeper than 0.5% are surprisingly rare outside hilly or mountainous areas.

Effect of Insufficient Gradient

Due to the normal cross slope and the interaction with grade, road sections with insufficient drainage gradient are few and short. Still, they account for an unacceptable number of skid accidents. These hot spots are found at the entrances and exits of banked curves, where the cross slope changes direction in order to create superelevation. As the outside edge of the curve is raised (or superelevated) to create the bank, it passes through a point where the cross slope is absolutely flat. If there is not enough longitudinal grade, water will collect at these spots. This takes place at the beginnings and ends of curves to the left in countries with right hand traffic, and curves to the right in countries with left hand traffic.

A large study in Sweden has shown that the lanes on the outside of the curve (outercurves) average five times more crashes than lanes on the inner side (innercurves). This finding can partially be explained by the entrances and exits of banked outercurves having insufficient DG.

Minimizing Insufficient Drainage Gradient

Roads should be designed so that sections where the cross slope change direction (and sign), are located where the road is going uphill or downhill. Otherwise the pavement will get an area with too little drainage gradient (< 0.5%), resulting in unacceptable skid accident risk.

When designing road curves in a flat landscape, it may be necessary to design long wave undulations on purpose. These "synthetic" longitudinal gradients can then be used to reach a sufficient drainage gradient, in sections where the cross slope is close to zero.

Another option to minimize crash risk due to low DG at entrance or exit of banked outercurves, is to move the superelevation further from the curve and out to a straight road section. This results in a banked straight lane. This design can yield another risk, since the water film (when raining) on an adjacent lane may become thicker. However, this is on a straight section where the lack of road curvature minimize the lateral forces and thus keep the skid risk low.

Yet an option is to – within the superelevation transition section – increase the cross slope "tilt rate" within the zone where the cross slope is between −0.5 to +0.5%.

Maintenance

As roads are worn down by tire wear, wheelpath rutting often interrupts the DG, and water will collect in the wheelpaths. This is more prevalent on asphalt pavement, but concrete pavements are not immune. Before wet pavement crashes reach unacceptable levels, maintenance actions such as resurfacing or diamond grinding should be taken, even if other pavement distress such as cracking is still low

Road Infrastructure

It consists of the installation of fixed assets including surface roads and railways and terminals such as bus stops, trucking terminals, railways stations. Operations of vehicles along these transport systems make the travel time to reduce and generation of employment in the sector which in aggregate influence the aggregate demand for goods and services that ultimately leads to increase in GDP and overall development.

Intersection (Road)

Overview

Intersection is an area shared by two or more roads. This area is designated for the vehicles to turn to different directions to reach their desired destinations. Its main function is to guide vehicles to their respective directions. Traffic intersections are complex locations on any highway. This is because vehicles moving in different direction wan to occupy same space at the same time. In addition, the pedestrians also seek same space for crossing. Drivers have to make split second decision at an intersection by considering his route, intersection geometry, speed and direction of other vehicles etc. A small error in judgment can cause severe accidents. It also causes delay and it depends on type, geometry, and type of control. Overall traffic flow depends on the performance of the intersections. It also affects the capacity of the road. Therefore, both from the accident perspective and the capacity perspective, the study of intersections very important for the traffic engineers especially in the case of urban scenario.

Conflicts at an Intersection

Conflicts at an intersection are different for different types of intersection. Consider a typical four-legged intersection as shown in figure. The number of conflicts for competing through movements are 4, while competing right turn and through movements are 8. The conflicts between right turn traffics are 4, and between left turn and merging traffic is 4. The conflicts created by pedestrians will be 8 taking into account all the four approaches. Diverging traffic also produces about 4 conflicts. Therefore, a typical four legged intersection has about 32 different types of conflicts.

The essence of the intersection control is to resolve these conflicts at the intersection for the safe and efficient movement of both vehicular traffic and pedestrians. Two methods of intersection controls are there: time sharing and space sharing. The type of intersection control that has to be adopted depends on the traffic volume, road geometry, cost involved, importance of the road etc.

Levels of Intersection Control

The control of an intersection can be exercised at different levels. They can be either passive control, semi control, or active control. In passive control, there is no explicit control on the driver . In semi control, some amount of control on the driver is there from the traffic agency. Active control means the movement of the traffic is fully controlled by the traffic agency and the drivers cannot simply maneuver the intersection according to his choice.

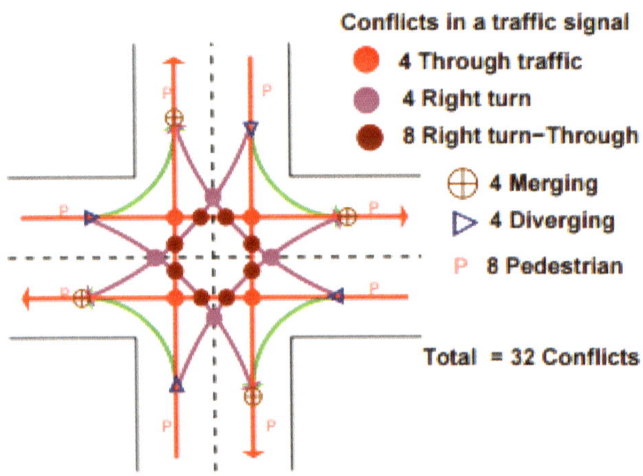

Figure: Conflicts at an intersection

Passive Control

When the volume of traffic is less, no explicit control is required. Here the road users are required to obey the basic rules of the road. Passive control like traffic signs, road markings etc. are used to complement the intersection control. Some of the intersection control that are classified under passive control are as follows:

1. No control If the traffic coming to an intersection is low, then by applying the basic rules of the road like driver on the left side of the road must yield and that through movements will have priority than turning movements. The driver is expected to obey these basic rules of the road.

2. Traffic signs: With the help of warning signs, guide signs etc. it is able to provide some level of control at an intersection. Give way control, two-way stop control, and all-way stop control are some examples. The GIVE WAY control requires the driver in the minor road to slow down to a minimum speed and allow the vehicle on the major road to proceed. Two way stop control requires the vehicle drivers on the minor streets should see that the conflicts are avoided. Finally an all-way stop control is usually used when it is difficult to differentiate between the major and minor roads in an intersection. In such a case, STOP sign is placed on all the approaches to the intersection and the driver on all the approaches are required to stop the vehicle. The vehicle at the right side will get priority over the left approach. The traffic control at 'at-grade' intersection may be uncontrolled in cases of low traffic. Here the road users are required to obey the basic rules of the road. Passive control like traffic signs, road markings etc. are used to complement the intersection control.

3. Traffic signs plus marking: In addition to the traffic signs, road markings also complement the traffic control at intersections. Some of the examples include stop line marking, yield lines, arrow marking etc.

Semi Control

In semi control or partial control, the drivers are gently guided to avoid conflicts. Channelization and traffic rotaries are two examples of this.

1. Channelization: The traffic is separated to flow through definite paths by raising a portion of the road in the middle usually called as islands distinguished by road markings. The conflicts in traffic movements are reduced to a great extent in such a case. In channelized intersections, as the name suggests, the traffic is directed to flow through different channels and this physical separation is made possible with the help of some barriers in the road like traffic islands, road markings etc.

2. Traffic rotaries: It is a form of intersection control in which the traffic is made to flow along one direction around a traffic island. The essential principle of this control is to convert all the severe conflicts like through and right turn conflicts into milder conflicts like merging, weaving and diverging. It is a form of 'at-grade' intersection laid out for the movement of traffic such that no through conflicts are there. Free-left turn is permitted where as through traffic and right-turn traffic is forced to move around the central island in a clock-wise direction in an orderly manner. Merging, weaving and diverging operations reduces the conflicting movements at the rotary.

Active Control

Active control implies that the road user will be forced to follow the path suggested by the traffic control agencies. He cannot maneuver according to his wish. Traffic signals and grade separated intersections come under this classification.

1. Traffic signals: Control using traffic signal is based on time sharing approach. At a given time, with the help of appropriate signals, certain traffic movements are restricted where as certain other movements are permitted to pass through the intersection. Two or more phases may be provided depending upon the traffic conditions of the intersection. When the vehicles traversing the intersection is very large, then the control is done with the help of signals. The phases provided for the signal may be two or more. If more than two phases are provided, then it is called multiphase signal.

The signals can operate in several modes. Most common are fixed time signals and vehicle actuated signals. In fixed time signals, the cycle time, phases and interval of each signal is fixed. Each cycle of the signal will be exactly like another. But they cannot cater to the needs of the fluctuating traffic. On the other hand, vehicle actuated signals can respond to dynamic traffic situations. Vehicle detectors will be placed on the streets approaching the intersection and the detector will sense the presence of the vehicle and pass the information to a controller. The controller then sets the cycle time and adjusts the phase lengths according to the prevailing traffic conditions.

2. Grade separated intersections: The intersections are of two types. They are at-grade intersections and grade-separated intersections. In at-grade intersections, all roadways join or cross at the same vertical level. Grade separated intersections allows the traffic to cross at different vertical levels. Sometimes the topography itself may be helpful in constructing such intersections. Otherwise, the initial construction cost required will be very high. Therefore, they are usually constructed on high speed facilities like expressways, freeways etc. These type of intersection increases the road capacity because vehicles can flow with high speed and accident potential is also reduced due to vertical separation of traffic.

Types

Road Segments

One way to classify intersections is by the number of road segments (arms) that are involved.

- A three-way intersection is a junction between three road segments (arms): a T junction when two arms form one road, or a Y junction – the latter also known as a fork if approached from the stem of the Y.
- A four-way intersection, or crossroads, usually involves a crossing over of two streets or roads. In areas where there are blocks and in some other cases, the crossing streets or roads are perpendicular to each other. However, two roads may cross at a different angle. In a few cases, the junction of two road segments may be offset from each when reaching an intersection, even though both ends may be considered the same street.
- Five-way intersections are less common but still exist, especially in urban areas with non-rectangular blocks. An example of this is the intersection for which the Five Points district in Atlanta is named.
- Six-way intersections usually involve a crossing of three streets at one junction; for example, a crossing of two perpendicular streets and a diagonal street is a rather common type of 6-way intersection.
- Seven or more approaches to a single intersection, such as at Seven Dials, London, are rare.

Intersection along the Veterans Memorial Parkway, an at-grade limited-access road in London, Ontario.

Traffic Controls

Another way of classifying intersections is by traffic control technology:

- Uncontrolled intersections, without signs or signals (or sometimes with a warning sign). Priority (right-of-way) rules may vary by country: on a 4-way intersection traffic from the right often has priority; on a 3-way intersection either traffic from the right has priority again, or traffic on the continuing road. For traffic coming from the same or opposite direction, that which goes straight has priority over that which turns off.
- Yield-controlled intersections may or may not have specific “YIELD” signs (known as “GIVE WAY” signs in some countries).
- Stop-controlled intersections have one or more “STOP” signs. Two-way stops are common, while some countries also employ four-way stops.
- Signal-controlled intersections depend on traffic signals, usually electric, which indicate which traffic is allowed to proceed at any particular time.

Lane Design

- A traffic circle is a type of intersection at which traffic streams are directed around a circle. Types of traffic circles include roundabouts, ‘mini-roundabouts’, ‘rotaries’, “STOP”-controlled circles, and signal-controlled circles. Some people consider roundabouts to be a distinct type of intersection from traffic circles (with the distinction based on certain differences in size and engineering).
- A box junction can be added to an intersection, generally prohibiting entry to the intersection unless the exit is clear.
- Some (unconventional or alternative) intersections employ indirect left turns to increase capacity and reduce delays. The Michigan left combines a right turn and a U-turn. Jughandle lefts diverge to the right, then curve to the left, converting a left turn to a crossing maneuver, similar to throughabouts. These techniques are generally used in conjunction with signal-controlled intersections, although they may also be used at stop-controlled intersections.

- Other designs include advanced stop lines, parallel-flow and continuous-flow intersections, hook turns, quadrants, seagull intersections, slip lanes, staggered junctions, superstreets, Texas Ts, Texas U-turns and turnarounds.
- A roundabout and its variants like turbo roundabouts, bowties and distributing circles like traffic circles and right-in/right-out (RIRO) intersections.

Turns

At intersections, turns are usually allowed, but are often regulated to avoid interference with other traffic. Certain turns may be not allowed or may be limited by regulatory signs or signals, particularly those that cross oncoming traffic. Alternative designs often attempt to reduce or eliminate such potential conflicts.

Turn Lanes

At intersections with large proportions of turning traffic, turn lanes (also known as turn bays) may be provided.

Turn lanes allow vehicles to cross oncoming traffic (i.e., a left turn in right-side driving countries, or a right turn in left-side driving countries), or to exit a road without crossing traffic (i.e., a right turn in right-side driving countries, or a left turn in left-side driving countries). Absence of a turn lane does not normally indicate a prohibition of turns in that direction. Instead, traffic control signs are used to prohibit specific turns.

Turn lanes can increase the capacity of an intersection and/or improve safety. Turn lanes can have a dramatic effect on the safety of a junction. In rural areas, crash frequency can be reduced by up to 48% if left turn lanes are provided on both main-road approaches at stop-controlled intersections. At signalized intersections, crashes can be reduced by 33%. Results are slightly lower in urban areas.

Turn lanes are marked with an arrow bending into the direction of the turn which is to be made from that lane. Multi-headed arrows indicate that vehicle drivers may travel in any one of the directions pointed to by an arrow.

Turn Signals

Traffic signals facing vehicles in turn lanes often have arrow-shaped indications. Green arrows indicate protected turn phases, when vehicles may turn unhindered by oncoming traffic. Red arrows may be displayed to prohibit turns in that direction. Red arrows may be displayed along with a circular green indication to show that turns in the direction of the arrow are prohibited, but other movements are allowed. In some jurisdictions, a red arrow prohibits a turn on red.

Disadvantages to turn lanes include increased pavement area, with associated increases in construction and maintenance costs, as well as increased amounts of stormwater runoff. They also increase the distance over which pedestrians crossing the street are exposed to vehicle traffic. If a turn lane has a separate signal phase, it often increases the delay experienced by oncoming through traffic. Without a separate phase, left crossing traffic does not get the full safety benefit of the turn lane.

Lane management

Alternative intersection configurations can manage turning traffic to increase safety and intersection throughput. These include the Michigan left, "superstreet" and continuous flow intersection.

Diagram of an example intersection of two-way streets as seen from above (traffic flows on the right side of the road). The East-West street has left turn lanes from both directions, but the North-South street does not have left turn lanes at this intersection. The East-West street traffic lights also have green left turn arrows to show when unhindered left turns can be made. Some possible markings for crosswalks are shown as examples.

Pedestrians

Times Square is the hub of the Broadway theater district and a major cultural venue in Midtown Manhattan, New York City. The pedestrian intersection also has one of the highest annual attendance rates of any tourist attraction in the world, estimated at 60 million.

Intersections generally must manage pedestrian as well as vehicle traffic. Pedestrian aids include crosswalks, pedestrian-directed traffic signals ("walk light") and over/underpasses. Walk lights may be accompanied by audio signals to aid the visually impaired. Medians can offer pedestrian islands, allowing pedestrians to divide their crossings into a separate segment for each traffic direction, possibly with a separate signal for each.

Some intersections display red lights in a all directions for a period of time. Known as a pedestrian scramble, this type of vehicle all-way stop allows pedestrians to cross safely in any direction, including diagonally. All green for non motorists is known from the crossing at Shibuya Station, Tokyo.

Protected Intersections for Bicycles

A number of features make this protected intersection much safer. A corner refuge island, a setback crossing of the pedestrians and cyclists, generally between 1.5–7 metres of setback, a forward stop bar, which allows cyclists to stop for a traffic light well ahead of motor traffic who must stop behind the crosswalk. Separate signal staging or at least an advance green for cyclists and pedestrians is used to give cyclists and pedestrians no conflicts or a head start over traffic. The design makes a right turn on red, and sometimes left on red depending on the geometry of the intersection in question, possible in many cases, often without stopping.

Cyclists ideally have a protected bike lane on the approach to the intersection, separated by a concrete median with splay kerbs if possible, and have a protected bike lane width of at least 2 metres if possible (one way). In the Netherlands, most one way cycle paths are at least 2.5 metres wide.

Intersections can be made more bicycle-friendly with various interventions such as protected intersections or grade separation. A low grade alternative is a bike box in terms of safety.

Lane

Thru lanes indicated by arrows on California CR G4 (Montague Expressway) in Silicon Valley.

In the context of traffic control, a lane is part of a roadway (carriageway) that is designated for use by a single line of vehicles, to control and guide drivers and reduce traffic conflicts. Most public roads (highways) have at least two lanes, one for traffic in each direction, separated by lane markings. On multilane roadways and busier two-lane roads, lanes are designated with road surface markings. Major highways often have two multi-lane roadways separated by a median.

Some roads and bridges that carry very low volumes of traffic are less than 15 feet (4.6 m) wide, and are only a single lane wide. Vehicles travelling in opposite directions must slow or stop to pass each other. In rural areas, these are often called country lanes. In urban areas, alleys are often only one lane wide. Urban and suburban one lane roads are often designated for one-way traffic.

Capacity

Lane capacity varies widely due to conditions such as neighboring lanes, lane width, elements next to the road, number of driveways, presence of parking, speed limits, number of heavy vehicles and so on – the range can be as low as 1000 passenger cars / hour to as high as 4800 passenger cars / hour but mostly falls between 1500 and 2400 passenger cars / hour.

Types

The Ontario Highway 401 in the Greater Toronto area, with 17 travel lanes in 6 separate carriageways visible in the midground.

Turning lane on the Rodovia BR-101 (Brazil)

Transfer lanes, connecting surface collector lanes with through lanes between two tunnels

A left-turn merging lane in Germany, needing explanation by a crafted sign

- A traffic lane or travel lane is a lane for the movement of vehicles traveling from one destination to another, not including shoulders.
- A through lane or thru lane is a traffic lane for through traffic. At intersections, these may be indicated by arrows on the pavement pointing straight ahead.

- An express lane of a road is used for faster moving traffic and has less access to exits/off ramps. In other areas, an express lane may refer to a HOV lane.
- A reversible lane (contraflow lane) is a lane where the direction of traffic can be changed to match the peak flow. They are used to accommodate periods of high traffic flow, especially rush hour where the flow is predominantly in one direction, on roads that cannot be easily widened such as over bridges. One or more lanes are removed from the opposing flow and added to the peak flow.
- An auxiliary lane is a lane other than a through lane, used to separate entering, exiting or turning traffic from the through traffic.
    - An acceleration lane or merge lane allows traffic entering a highway to accelerate to the speed of through traffic before merging with it.
    - A deceleration lane is a lane adjacent to the primary road or street used to improve traffic safety by allowing drivers to pull out of the through lane and decelerate safely before turning off a surface street or exiting a freeway.
    - A turn lane is set aside for slowing down and making a turn, so as not to disrupt traffic. By removing turning traffic from the through lanes, motorist safety is improved and delay is removed, but crossing distances for pedestrians are lengthened, increasing their exposure to collisions.
    - A two-way center turn lane is a lane in the center of a roadway to allow drivers traveling in either direction to pause before turning across oncoming traffic (a left turn in right-driving countries, or a right turn in left-driving countries) while waiting for a gap in traffic.
    - A passing lane is sometimes provided on busy two-lane roads to allow drivers to pass slower vehicles without having to cross the center line and risking a head-on collision.
- A climbing lane, truck lane, or crawler lane is often provided on steep mountain grades, in order to allow smaller vehicles to pass larger, slower ones. The lane is marked only on the uphill stretch and usually a short distance afterward (for regaining speed).
- An operational lane or auxiliary lane is an extra lane on the entire length of highway between interchanges, giving drivers more time to merge in or out. The lane is created when an entrance ramp meets the highway, and drops out (with an "exit only" sign) to become the ramp at the next exit.
- A transfer lane of a road is used to move from express lanes to collector lanes, or vice versa; it is somewhat similar to an auxiliary lane.
- A collector lane of a road is used for slower moving traffic and has more access to exits/off ramps.
- Dedicated lanes are traffic lanes set aside for particular types of vehicles.
    - A high occupancy vehicle or carpool lane is reserved for carpooling. In the US, they may be marked with a diamond icon every few hundred feet (hence the nickname "diamond

lane"), or separated from other lanes by double broken white lines, a continuous pair of double yellow lines, or just a single broken white line.

- A high-occupancy toll lane is a combination of an HOV lane and toll collection technology that allows drivers without passengers to use the HOV lane by paying a premium price for the privilege
- A designated bicycle lane is a portion of the roadway or shoulder designated for the exclusive or preferential use of bicyclists. This designation is indicated by special word and/or symbol markings on the pavement and "BIKE LANE" signs.
- A motorcycle lane is provided at certain roads and highways such as the Federal Highway in Malaysia to segregate the motorcycle traffic from the main roadways to reduce motorcycle-related accidents. The motorcycle lane may form a part of the hard shoulder, or may be one or more completely separated lanes.
- A bus lane is reserved for buses providing public transportation on a fixed route, sometimes with overhead catenary for trolleybuses. In some countries, bus lanes may also be used by some other traffic, such as taxis, bicycles and motorbikes.
- A tram lane is a lane reserved for the use of buses, trams and taxicabs. It is usually encountered in cities with curbside tram network, such as Zagreb.
- A truckway is a dedicated lane for longer length trucks; for instance, the Florida Turnpike allows 96-foot long double trailer combinations, in contrast to normal Florida highways' 53-foot limit. Since the major cost of trucking is the fixed cost of the same trailer with its driver the cost per ton of operating with truckway size and weight allowances is 35 to 40 percent below the cost of operations on the non-truckways.

- In some areas, the lane adjacent to the curb is reserved for non-moving vehicles.
  - A parking lane is reserved for parallel parking of vehicles.
  - A fire lane is the area next to a curb, which is reserved for firefighting equipment, ambulances, or other emergency vehicles. Parking in these areas, often marked by red lines, usually warrants a parking ticket.
  - A loading lane (loading zone in the United States) is an area next to a curb, which is reserved for loading and unloading passengers and/or freight. It may be marked by a sign ("LOADING ONLY" or "LOADING ZONE") or by a yellow or white-painted curb.
  - A hard shoulder is sometimes called an emergency lane or a breakdown lane, when it is reserved for vehicle breakdowns, and for emergency vehicles. On some congested roads, the shoulder is used as a vehicle lane during peak travel hours.
- By law or custom, inside lanes are often reserved for faster traffic, and outside lanes are used by slower traffic.
  - An overtaking lane is the lane furthest from the shoulder of a multi-lane carriageway/roadway (sometimes called the fast lane, although this is deprecated by the authorities), i.e. the leftmost lane where traffic drives on the right, or the rightmost lane where traffic drives in the left.

- The slow lane is the lane nearest to the shoulder of a multi-lane carriageway/roadway, i.e. the farthest from the overtaking lane.

These usages lead to the phrases *life in the slow lane* and *life in the fast lane*, used to describe relaxed or busy lifestyles, respectively and used as the titles of various books and songs.

Lane Markings

A typical rural American freeway (Interstate 5 in the Central Valley of California). Notice the yellow line on the left, the dashed white line in the middle, and the solid white line on the right. Also note the rumble strip to the left of the yellow line.

Painted lane markings vary widely from country to country. In the United States, Canada, Mexico, Honduras, Puerto Rico, Virgin Islands and Norway, yellow lines separate traffic going opposite directions and white separates lanes of traffic traveling the same direction, but such is not the case in many European countries.

Lane markings are mostly lines painted on the road by a road marking machine, which can adjust the marking widths according to the lane type.

Numbering of Freeway Lanes in California

Traffic reports in California often refer to accidents being "in the number X lane." The California Department of Transportation (Caltrans) assigns the numbers from left to right. The far left passing lane is the number 1 lane. The number of the slow lane (closest to freeway onramps/offramps) depends on the total number of lanes, and could be anywhere from 2 to 8.

Junction (Road)

Junctions, at-grade or grade separated, are locations of high accident concentration. In most countries 40 - 60% of the total number of accidents occurs at junctions. Consequently, special attention should be given in determining the type, the shape of junctions, as well as the number of junctions along a road axis and the efficient design of each one.

Junctions vary from quiet unmarked junctions to busy multi-lane box junctions that are controlled by traffic lights. Examiners will include as many of these varied junction during your test.

Understanding exactly how to deal with each and every type of junction is not only essential for safe long-term driving, but also necessary for passing the practical driving test.

Road Junctions are designed at places where vehicular traffic can move in different directions in a systematic way. Road Junctions decrease the probability of accidents. The traffic at Road junctions is controlled and resumed in a systematic way to proceed further in their respective directions with the help of traffic signals.

Road Junction in Taiwan

Design of Road Junctions is a crucial subject. Understanding the nature of traffic, the kind of area, density of population etc is very important so as to propose a suitable road Junction design.

Types

The most basic distinction among junction types is whether or not the roads cross at the same or different elevations. More expensive, grade-separated *interchanges* generally offer higher throughput at higher cost. Single-grade *intersections* are lower cost and lower throughput. Each main type comes in many variants.

Interchange

At interchanges, roads pass above or below each other, using grade separation and slip roads. The terms *motorway junction* and *highway junction* typically refer to this layout. They can be further subdivided into those with and without signal controls.

Signalized (traffic-light controlled) interchanges include such "diamond" designs as the diverging diamond, Michigan urban diamond, three-level diamond, and tight diamond. Others include center-turn overpass, contraflow left, single loop, and single-point urban overpass.

Non signalized designs include the cloverleaf, contraflow left, dogbone (restricted dumbbell), double crossover merging, dumbbell (grade-separated bowtie), echelon, free-flow interchange, partial cloverleaf, raindrop, single and double roundabouts (grade-separated roundabout), single-point urban, stack, and windmill.

Autobahnkreuz (literally autobahn cross), short form *kreuz*, and abbreviated as AK, is a four-way interchange on the German autobahn network. *Autobahndreieck* (literally autobahn triangle), short form *dreieck*, and abbreviated as AD, is a three-way interchange on the German autobahn network.

Intersection

At intersections, roads cross at-grade. They also can be further subdivided into those with and without signal controls.

Signalized designs include advanced stop line, bowtie, box junction, continuous-flow intersection, continuous Green-T, double-wide, hook turn, jughandle, median u-turn, Michigan left, paired, quadrant, seagulls, slip lane, split, staggered, superstreet, Texas T, Texas U-turn and turnarounds.

Non-signalized designs include unsignalized variations on continuous-flow 3 and 4-leg, median u-turn and superstreet, along with Maryland T/J, roundabout and traffic circle.

Pedestrians

Intersections generally must manage pedestrian as well as vehicle traffic. Pedestrian aids include crosswalks, pedestrian-directed traffic signals ("walk light") and over/underpasses. Walk lights may be accompanied by audio signals to aid the visually impaired. Medians can offer pedestrian islands, allowing pedestrians to divide their crossings into a separate segment for each traffic direction, possibly with a separate signal for each.

Protected Intersections for Bicycles

A number of features make this protected intersection much safer. A corner refuge island, a setback crossing of the pedestrians and cyclists, generally between 1.5–7 metres of setback, a forward stop bar, which allows cyclists to stop for a traffic light well ahead of motor traffic who must stop behind the crosswalk. Separate signal staging or at least an advance green for cyclists and pedestrians is used to give cyclists and pedestrians no conflicts or a head start over traffic. The design makes a right turn on red, and sometimes left on red depending on the geometry of the intersection in question, possible in many cases, often without stopping.

Cyclists ideally have a protected bike lane on the approach to the intersection, separated by a concrete median with splay kerbs if possible, and have a protected bike lane width of at least 2 metres if possible (one way). In the Netherlands, most one way cycle paths are at least 2.5 metres wide.

Bicycle traffic can be accommodated with the low grade bike lanes in the roadway or higher grade and much safer Protected Bicycle Path that are physically separated from the roadway.

Interchange (Road)

An interchange is a road junction where two roads are connected by dedicated roadways, called ramps. The roads connected by an interchange do not intersect one another directly, and if they cross, the crossing is grade-separated.

Since interchanges often involve grade-separated crossings, the road elevation of the segments becomes important. If two roads cross without connecting directly, their elevations must be different.

Sometimes you will find interchanges which are intersection of two roads, where one road is for exiting and the other road is for entering the freeway. Negotiating these interchanges requires cooperation and communication. Cars exiting have the right-of-way unless there is a safety threat.

Terminology

Note: The descriptions of road junctions are for countries where vehicles drive on the right side of the road. For countries where driving is on the left the layout of the junctions is the same, only left/right is reversed.

- A *freeway junction* or *highway interchange* (in the US) or *motorway junction* (in the UK) is a type of *road junction*, linking one controlled-access highway (freeway or motorway) to another; to other roads; or to a rest area/motorway service area. On the UK motorway network, most (but not all) junctions with other roads are numbered sequentially. In the U.S., interchanges are either numbered according to cardinal interchange number, or by mileage (typically the latter in most states).
- A *highway ramp* (as in *exit ramp / off-ramp* and *entrance ramp / on-ramp*) (North American usage) or *slip road* is a short section of road which allows vehicles to enter or exit a controlled-access highway (freeway or motorway).
- A *directional* ramp always tends toward the desired direction of travel. This means that a ramp that makes a left turn exits from the left side of the roadway (a left exit). Left directional ramps are relatively uncommon as the left lane is usually reserved for high-speed through traffic. Ramps for a right turn are almost always right directional ramps. Where traffic drives on the left, these cases are reversed.
- A *non-directional* ramp goes in a direction opposite to the desired direction of travel. Many loop ramps (as in a cloverleaf) are non-directional.
- A *semi-directional* ramp exits a road in a direction opposite from the desired direction of travel, but then turns toward the desired direction of travel. Many flyover ramps (as in a stack) are semi-directional.
- A *U-turn* ramp leaves the road in one driving direction, turns over or under it and rejoins in the opposite direction.

Off-ramp accessed from collector lanes along Highway 401 in Toronto, Ontario, Canada.

Weaving is an undesirable situation in which traffic veering right and traffic veering left must cross paths within a limited distance, to merge with traffic on the through lane.

The German Autobahn system splits Autobahn-to-Autobahn interchanges into two types: a four-way interchange, the *Autobahnkreuz* (AK), where two motorways cross, and a three-way interchange, the *Autobahndreieck* (AD) where two motorways merge.

Some on-ramps have a ramp meter, which is a dedicated ramp-only traffic light that throttles the flow of entering vehicles.

Complete and Incomplete Interchanges

A *complete interchange* has enough ramps to provide access from any direction of any road in the junction to any direction of any other road in the junction. A complete interchange between a freeway and another road (not a freeway) requires at least four ramps. Complete interchanges between two freeways generally have at least eight ramps, as having fewer would considerably reduce capacity and increase weaving. Using U-turns, the number for two freeways can be reduced to six, by making cars that want to turn left either pass by the other road first, then make a U-turn and turn right, or turn right first and then make a U-turn. Depending on the interchange type and the connectivity offered other numbers of ramps may be used. For example, if a highway interchanges with a highway containing a collector/express system, additional ramps can be used to strictly link the interchanging highway with the collector and express lanes respectively. For highways with high-occupancy vehicle (HOV) lanes, ramps can be used to service these carriageways directly, thereby increasing the number of ramps used.

An *incomplete interchange* has at least one or more missing ramps that prevent access to at least one direction of another road in the junction from any other road in the junction.

Cloverleaf Interchange

A cloverleaf interchange allows for interchange of two freeways with minimal disruption of speed or movement. The cloverleaf usually has characteristic weave lanes as traffic enters from one roadway and exits from another. Curved exits lead to braking and steering problems as drivers adjust from high speed to the speed of a curve.

Diamond Interchange

The diamond interchange allows for interchange of a major roadway with a secondary dual or multi-lane roadway. The diamond interchange may have traffic control devices on the intersecting secondary roadway that allows for left and right turns onto the secondary roadway.

Trumpet Interchange

The trumpet interchange allows for interchange of secondary two-way streets to a multi-lane roadway with minimal traffic mix. It takes the place of a T-intersection which is often used when a roadway ends at the intersection of another roadway.

Frontage Road Interchange

Frontage road interchanges allow for interchange of vehicles using parallel secondary two-way or one-way roadways and a major multi-lanc roadway. Frontage road turnarounds allow drivers to exit a multi-lane roadway and use the opposing frontage road to enter the multi-lane roadway in the opposite direction.

Channelized Intersection

Vehicles approaching an intersection are directed to definite paths by islands, marking etc. and this method of control is called channelization. Channelized intersection provides more safety and efficiency. It reduces the number of possible conflicts by reducing the area of conflicts available in the carriageway. If no channelizing is provided the driver will have less tendency to reduce the speed while entering the intersection from the carriageway. The presence of traffic islands, markings etc. forces the driver to reduce the speed and becomes more cautious while maneuvering the intersection. A channelizing island also serves as a refuge for pedestrians and makes pedestrian crossing safer. Channelization of traffic through a three-legged intersection and a four-legged intersection.

Summary

Traffic intersections are problem spots on any highway, which contribute to a large share of accidents. For safe operation, these locations should be kept under some level of control depending upon the traffic quantity and behavior. Based on this, intersections and interchnges are constructed,

References

- O'Flaherty, edited by C.A. (2002). Highways the location, design, construction and maintenance of road pavements (4th ed.). Oxford: Butterworth-Heinemann. ISBN 978-0-7506-5090-8
- Unger, N; Bond, T. C; Wang, J. S; Koch, D. M; Menon, S; Shindell, D. T; Bauer, S (2010). "Attribution of climate forcing to economic sectors". Proceedings of the National Academy of Sciences. 107(8): 3382–87. doi:10.1073/pnas.0906548107. PMC 2816198 . Retrieved 2013-08-14
- "The Rebecca Riots". Rebecca and her daughters come to Rhayader. Victorian Powys for Schools. March 2002. Retrieved 2007-04-28
- "Traffic-related Air Pollution near Busy Roads". American Journal of Respiratory and Critical Care Medicine Vol 170. pp. 520–26. 2004
- Lay, Maxwell G (1992). Ways of the World: A History of the World›s Roads and of the Vehicles that Used Them. Rutgers University Press. ISBN 0-8135-2691-4
- "Road Infrastructure Strategic Framework for South Africa; A Discussion Document". National Department of Transport (South Africa). Archived from the original on 2007-09-27. Retrieved 2007-03-24
- Helbing, D.; Molnär, P.; Farkas, I.J.; Bolay, K. Environment and Planning B: Planning and Design 2001, volume 28 (Self-organizing pedestrian movement) (PDF). p. 376. doi:10.1068/b2697

4

Traffic Control and Management

Traffic control and management deals with the planning and control of transport services for the effective movement of vehicles and freight. In order to completely understand traffic control and management, it is necessary to understand the techniques and methods related to it. The following chapter elucidates the varied aspects of traffic flow, traffic congestion, traffic reporting, road traffic control, traffic optimization, active traffic management, etc.

Traffic

Traffic and transportation are of essential importance for the global economy and society. Growing need of transportation and limitation of fossil fuel imply that economical use of transportation resources is critical and will become increasingly important in the future. Simultaneously enterprises are forced to reduce costs and save resources to increase their competitiveness in the global markets. Optimization models and computational solution methods have been developed to manage and control complex systems arising in traffic, transportation and logistics. However, such problems often become very difficult and cannot be solved to optimality with today's technologies. New models and methods are needed to be able to compute better solutions for practical use. The modeling and solution approaches are often based on mathematical programming and complemented with heuristic methods.

Traffic is the movement of people and goods from one location to another. The movement typically occurs along a specific facility or pathway that can be called a guideway. It may be a physical guideway, as in the case of a railroad, or it may be an agreed-upon or designated route, marked either electronically (as in aviation) or geographically (as in the maritime industry). Movement—excepting pedestrian movement, which only requires human power—involves a vehicle of some type that can serve for people, goods, or both. Vehicle types, often referred to as modes of transportation, can be broadly characterized as road, rail, air, and maritime (*i.e.,* water-based).

Traffic evolves because of a need to move people and goods from one location to another. As such, the movement is initiated because of decisions made by people to transport themselves or others from one location to another to participate in activities at that second location or to move goods to a location where they have higher value. Traffic flows thus differ fundamentally from other areas of engineering and the physical sciences (such as the movement of electrons in a wire), because they are primarily governed and determined by laws of human behaviour. While physical attributes are critical in the operation of all modes (*e.g.,* to keep airplanes in the air), the demand or need to travel that gives rise to traffic is derived from the desire to change locations.

Rules of the Road

Rules of the road and driving etiquette are the general practices and procedures that road users are required to follow. These rules usually apply to all road users, though they are of special importance to motorists and cyclists. These rules govern interactions between vehicles and with pedestrians. The basic traffic rules are defined by an international treaty under the authority of the United Nations, the 1968 *Vienna Convention on Road Traffic*. Not all countries are signatory to the convention and, even among signatories, local variations in practice may be found. There are also unwritten local rules of the road, which are generally understood by local drivers.

Traffic controller on Michigan Avenue in Chicago, Illinois.

Traffic control in Rome, Italy. This traffic control podium can retract back to road level when not in use.

As a general rule, drivers are expected to avoid a collision with another vehicle and pedestrians, regardless of whether or not the applicable rules of the road allow them to be where they happen to be.

In addition to the rules applicable by default, traffic signs and traffic lights must be obeyed, and instructions may be given by a police officer, either routinely (on a busy crossing instead of traffic lights) or as road traffic control around a construction zone, accident, or other road disruption.

These rules should be distinguished from the mechanical procedures required to operate one's vehicle.

Directionality

Traffic going in opposite directions should be separated in such a way that they do not block each other's way. The most basic rule is whether to use the left or right side of the road.

Traffic Regulations

In many countries, the rules of the road are codified, setting out the legal requirements and punishments for breaking them.

In the United Kingdom, the rules are set out in the Highway Code, which includes not only obligations but also advice on how to drive sensibly and safely.

In the United States, traffic laws are regulated by the states and municipalities through their respective traffic codes. Most of these are based at least in part on the Uniform Vehicle Code, but there are variations from state to state. In states such as Florida, traffic law and criminal law are separate, therefore, unless someone flees a scene of an accident, commits vehicular homicide or manslaughter, they are only guilty of a minor traffic offense. However, states such as South Carolina have completely criminalized their traffic law, so, for example, one is guilty of a misdemeanor simply for travelling 5 miles over the speed limit.

Organized Traffic

Priority (Right of Way)

Vehicles often come into conflict with other vehicles and pedestrians because their intended courses of travel intersect, and thus interfere with each other's routes. The general principle that establishes who has the right to go first is called "right of way", or "priority". It establishes who has the right to use the conflicting part of the road and who has to wait until the other does so.

Yield to military traffic

Signs, signals, markings and other features are often used to make priority explicit. Some signs, such as the stop sign, are nearly universal. When there are no signs or markings, different rules are observed depending on the location. These default priority rules differ between countries, and may even vary within countries. Trends toward uniformity are exemplified at an international level by the Vienna Convention on Road Signs and Signals, which prescribes standardized traffic control devices (signs, signals, and markings) for establishing the right of way where necessary.

Crosswalks (or pedestrian crossings) are common in populated areas, and may indicate that pedestrians have priority over vehicular traffic. In most modern cities, the traffic signal is used to establish the right of way on the busy roads. Its primary purpose is to give each road a duration of time in which its traffic may use the intersection in an organized way. The intervals of time assigned for each road may be adjusted to take into account factors such as difference in volume of traffic, the needs of pedestrians, or other traffic signals. Pedestrian crossings may be located near other traffic control devices; if they are not also regulated in some way, vehicles must give priority to them when in use. Traffic on a public road usually has priority over other traffic such as traffic emerging from private access; rail crossings and drawbridges are typical exceptions.

Uncontrolled Traffic

Uncontrolled traffic comes in the absence of lane markings and traffic control signals. On roads without marked lanes, drivers tend to keep to the appropriate side if the road is wide enough. Drivers frequently overtake others. Obstructions are common.

Intersections have no signals or signage, and a particular road at a busy intersection may be dominant – that is, its traffic flows – until a break in traffic, at which time the dominance shifts to the other road where vehicles are queued. At the intersection of two perpendicular roads, a traffic jam may result if four vehicles face each other side-on.

Turning

Drivers often seek to turn onto another road or onto private property. The vehicle's blinking turn signals (commonly known as "blinkers" or "indicators") are often used as a way to announce one's intention to turn, thus alerting other drivers. The actual usage of directional signals varies greatly amongst countries, although its purpose is to indicate a driver's intention to depart from the current (and natural) flow of traffic well before the departure is executed (typically 3 seconds as a guideline).

Center turn lane on a Florida road

This will usually mean that turning traffic must stop and wait for a breach to turn, and this might cause inconvenience for drivers that follow them but do not want to turn. This is why dedicated lanes and protected traffic signals for turning are sometimes provided. On busier intersections where a protected lane would be ineffective or cannot be built, turning may be entirely prohibited, and drivers will be required to “drive around the block” in order to accomplish the turn. Many cities employ this tactic quite often; in San Francisco, due to its common practice, making three right turns is known colloquially as a “San Francisco left turn”. Likewise, as many intersections in Taipei City are too busy to allow direct left turns, signs often direct drivers to drive around the block to turn.

Turning rules are by no means universal. For example, in New Zealand (a drive-on-the-left country) between 1977 and 2012, left turning traffic had to give way to opposing right-turning traffic wishing to take the same road (unless there were multiple lanes, but then one must take care in case a vehicle jumped lanes). New Zealand abolished this particular rule on 25 March 2012, except at roundabouts or when denoted by a Give Way or Stop sign. Although the rule caused initial driver confusion, and many intersections required or still require modification, the change is predicted to eventually prevent one death and 13 serious injuries annually.

On roads with multiple lanes, turning traffic is generally expected to move to the lane closest to the direction they wish to turn. For example, traffic intending to turn right will usually move to the rightmost lane before the intersection. Likewise, left-turning traffic will move to the leftmost lane. Exceptions to this rule may exist where for example the traffic authority decides that the two rightmost lanes will be for turning right, in which case drivers may take whichever of them to turn. Traffic may adapt to informal patterns that rise naturally rather than by force of authority. For example, it is common for drivers to observe (and trust) the turn signals used by other drivers in order to make turns from other lanes. If several vehicles on the right lane are all turning right, a vehicle may come from the next-to-right lane and turn right as well, in parallel with the other right-turning vehicles.

Intersections

This intersection in San Jose, California has crosswalks, left-turn lanes, and traffic lights.

In most of Continental Europe, the default rule is to give priority to the right, but this may be overridden by signs or road markings. There, priority was initially given according to the social rank of each traveler, but early in the life of the automobile this rule was deemed impractical and replaced with the *priorité à droite* (priority to the right) rule, which still applies. At a traffic circle where *priorité à droite* is not overridden, traffic on what would otherwise be a roundabout gives way to traffic entering the circle. Most French roundabouts now have give-way signs for traffic entering the circle, but there remain some notable exceptions that operate on the old rule, such as the Place de l'Étoile around the Arc de Triomphe. Priority to the right where used in continental Europe may be overridden by an ascending hierarchy of markings, signs, signals, and authorized persons.

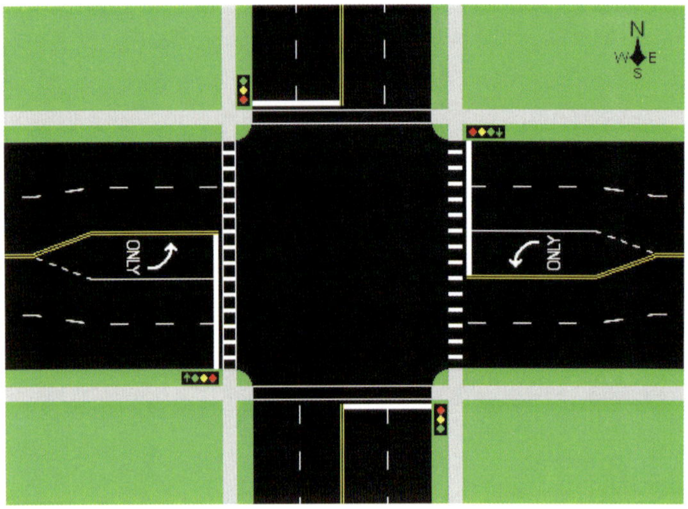

intersection of two-way streets as seen from above (traffic flows on the right side of the road). The East-West street has left turn lanes from both directions, but the North-South street does not have left turn lanes at this intersection. The East-West street traffic lights also have green left turn arrows to show when unhindered left turns can be made. Some possible markings for crosswalks are shown.

In the United Kingdom, priority is generally indicated by signs or markings, so that almost all junctions between public roads (except those governed by traffic signals) have a concept of a major road and minor road. The default give-way-to-the-right rule used in Continental Europe causes problems for many British and Irish drivers who are accustomed to having right of way by default unless otherwise indicated. A very small proportion of low-traffic junctions are unmarked – typically on housing estates or in rural areas. Here the rule is to "proceed with great care" i.e. slow the vehicle and check for traffic on the intersecting road.

Other countries use various methods similar to the above examples to establish the right of way at intersections. For example, in most of the United States, the default priority is to yield to traffic from the right, but this is usually overridden by traffic control devices or other rules, like the boulevard rule. This rule holds that traffic entering a major road from a smaller road or alley must yield to the traffic of the busier road, but signs are often still posted. The boulevard rule can be compared with the above concept of a major and minor road, or the priority roads that may be found in countries that are parties to the Vienna Convention on Road Signs and Signals.

Perpendicular intersections Also known as a "four-way" intersection, this intersection is the most common configuration for roads that cross each other, and the most basic type.

If traffic signals do not control a four-way intersection, signs or other features are typically used to control movements and make clear priorities. The most common arrangement is to indicate that one road has priority over the other, but there are complex cases where all traffic approaching an intersection must yield and may be required to stop.

In the United States, South Africa, and Canada, there are four-way intersections with a stop sign at every entrance, called four-way stops. A failed signal or a flashing red light is equivalent to a four-way stop, or an all-way stop. Special rules for four-way stops may include:

1. In the countries that use four-way stops, pedestrians always have priority at crosswalks – even at unmarked ones, which exist as the logical continuations of the sidewalks at every intersection with approximately right angles – unless signed or painted otherwise.
2. Whichever vehicle first stops at the stop line – or before the crosswalk, if there is no stop line – has priority.
3. If two vehicles stop at the same time, priority is given to the vehicle on the right.
4. If several vehicles arrive at the same time, a right-of-way conflict may arise wherein no driver has the legal right-of-way. This may result in drivers informally signaling to other drivers to indicate their intent to yield, for example by waving or flashing headlights.

In Europe and other places, there are similar intersections. These may be marked by special signs (according to the Vienna Convention on Road Signs and Signals), a danger sign with a black X representing a crossroads. This sign informs drivers that the intersection is uncontrolled and that default rules apply. In Europe and in many areas of North America the default rules that apply at uncontrolled four-way intersections are almost identical:

1. Rules for pedestrians differ by country, in the United States and Canada pedestrians generally have priority at such an intersection.
2. All vehicles must give priority to any traffic approaching from their right,
3. Then, if the vehicle is turning right or continuing on the same road it may proceed.
4. Vehicles turning left must also give priority to traffic approaching from the opposite direction, unless that traffic is also turning left.
5. If the intersection is congested, vehicles must alternate directions and/or circulate priority to the right one vehicle at a time.

Protected Intersection for Bicycles

A number of features make this protected intersection. A corner refuge island, a setback crossing of the pedestrians and cyclists, generally between 1.5–7 metres of setback, a forward stop bar, which allows cyclists to stop for a traffic light well ahead of motor traffic who must stop behind the crosswalk. Separate signal staging or at least an advance green for cyclists and pedestrians is used to give cyclists and pedestrians no conflicts or a head start over traffic. The design makes a right turn on red, and sometimes left on red depending on the geometry of the intersection in question, possible in many cases, often without stopping.

This type of intersection is common in the bicycle-friendly Netherlands.

Pedestrian Crossings

Pedestrians must often cross from one side of a road to the other, and in doing so may come into the way of vehicles traveling on the road. In many places pedestrians are entirely left to look after themselves, that is, they must observe the road and cross when they can see that no traffic will threaten them. Busier cities usually provide pedestrian crossings, which are strips of the road where pedestrians are expected to cross.

Slovenia

The actual appearance of pedestrian crossings varies greatly, but the two most common appearances are: (1) a series of lateral white stripes or (2) two longitudinal white lines. The former is usually preferred, as it stands out more conspicuously against the dark pavement.

Some pedestrian crossings accompany a traffic signal to make vehicles stop at regular intervals so pedestrians can cross. Some countries have "intelligent" pedestrian signals, where the pedestrian must push a button in order to assert his intention to cross. In some countries, approaching traffic is monitored by radar or by electromagnetic sensors buried in the road surface, and the pedestrian crossing lights are set to red if a speed infringement is detected. This has the effect of enforcing the local speed limit.

Pedestrian crossings without traffic signals are also common. In this case, the traffic laws usually states that the pedestrian has the right of way when crossing, and that vehicles must stop when a pedestrian uses the crossing. Countries and driving cultures vary greatly as to the extent to which this is respected. In the state of Nevada the car has the right of way when the crosswalk signal specifically forbids pedestrian crossing.

Some jurisdictions forbid crossing or using the road anywhere other than at crossings, termed *jaywalking*. In other areas, pedestrians may have the right to cross where they choose, and have right of way over vehicular traffic while crossing.

In most areas, an intersection is considered to have a crosswalk, even if not painted, as long as the roads meet at approximate right angles. The United Kingdom and Croatia are among the exceptions.

Pedestrian crossings may also be located away from intersections.

Level Crossings

A level crossing is an at-grade intersection of a railway by a road. Because of safety issues, they are often equipped with closable gates, crossing bells and warning signs.

An example of a typical rail crossing in the United States as an Amtrak Carolinian and Piedmont train passes through.

Speed limits

The higher the speed of a vehicle, the more difficult collision avoidance becomes and the greater the damage if a collision does occur. Therefore, many countries of the world limit the maximum speed allowed on their roads. Vehicles are not supposed to be driven at speeds which are higher than the posted maximum.

To enforce speed limits, two approaches are generally employed. In the United States, it is common for the police to patrol the streets and use special equipment (typically a radar unit) to measure the speed of vehicles, and pull over any vehicle found to be in violation of the speed limit. In Brazil, Colombia and some European countries, there are computerized speed-measuring devices spread throughout the city, which will automatically detect speeding drivers and take a photograph of the license plate (or number plate), which is later used for applying and mailing the ticket. Many jurisdictions in the U.S. use this technology as well.

A mechanism that was developed in Germany is the Grüne Welle, or *green wave*, which is an indicator that shows the optimal speed to travel for the synchronized green lights along that corridor. Driving faster or slower than the speed set by the behavior of the lights causes the driver to encounter many red lights. This discourages drivers from speeding or impeding the flow of traffic.

Overtaking

Overtaking (or *passing*) refers to a maneuver by which one or more vehicles traveling in the same direction are passed by another vehicle. On two-lane roads, when there is a split line or a dashed line on the side of the overtaker, drivers may overtake when it is safe. On multi-lane roads in most jurisdictions, overtaking is permitted in the "slower" lanes, though many require a special circumstance.

In the United Kingdom and Canada, notably on extra-urban roads, a solid white or yellow line closer to the driver is used to indicate that no overtaking is allowed in that lane. A double white or yellow line means that neither side may overtake.

In the United States, a solid white line means that lane changes are discouraged and a double white line means that the lane change is prohibited.

Lanes

When a street is wide enough to accommodate several vehicles traveling side-by-side, it is usual for traffic to organize itself into *lanes*, that is, parallel corridors of traffic. Some roads have one lane for each direction of travel and others have multiple lanes for each direction. Most countries apply pavement markings to clearly indicate the limits of each lane and the direction of travel that it must be used for. In other countries lanes have no markings at all and drivers follow them mostly by intuition rather than visual stimulus.

On roads that have multiple lanes going in the same direction, drivers may usually shift amongst lanes as they please, but they must do so in a way that does not cause inconvenience to other drivers. Driving cultures vary greatly on the issue of "lane ownership": in some countries, drivers traveling in a lane will be very protective of their right to travel in it while in others drivers will routinely expect other drivers to shift back and forth.

Designation and Overtaking

The usual designation for lanes on divided highways is the fastest lane is the one closest to the center of the road, and the slowest to the edge of the road. Drivers are usually expected to keep in the slowest lane unless overtaking, though with more traffic congestion all lanes are often used.

When driving on the left:

- The lane designated for faster traffic is on the right.
- The lane designated for slower traffic is on the left.
- Most freeway exits are on the left.
- Overtaking is permitted to the right, and sometimes to the left.

When driving on the right:

- The lane designated for faster traffic is on the left.
- The lane designated for slower traffic is on the right.
- Most freeway exits are on the right.
- Overtaking is permitted to the left, and sometimes to the right.

Countries party to the Vienna Convention on Road Traffic have uniform rules about overtaking and lane designation. The convention details (amongst other things) that "Every driver shall keep to the edge of the carriageway appropriate to the direction of traffic", and the "Drivers overtaking shall do so on the side opposite to that appropriate to the direction of traffic", notwithstanding the presence or absence of oncoming traffic. Allowed exceptions to these rules include turning or heavy traffic, traffic in lines, or situation in which signs or markings must dictate otherwise. These rules must be more strictly adhered to on roads with oncoming traffic, but still apply on multi-lane and divided highways. Many countries in Europe are party to the Vienna Conventions on traffic and roads. In Australia (which is not a contracting party), traveling in any lane other than the "slow" lane on a road with a speed limit at or above 80 km/h (50 mph) is an offence, unless signage is posted to the contrary or the driver is overtaking.

Many areas in North America do not have any laws about staying to the slowest lanes unless overtaking. In those areas, unlike many parts of Europe, traffic is allowed to overtake on any side, even in a slower lane. This practice is known as "passing on the right" in the United States and "overtaking on the inside" and "undertaking" in the United Kingdom. When referring to individual lanes on dual carriageways, one does not consider traffic travelling the opposite direction. The inside lane (in the British English sense, i.e. the lane beside the hard shoulder) refers to the lane used for normal travel, while the middle lane is used for overtaking cars on the inside lane. The outside lane (i.e. closest to oncoming traffic) is used for overtaking vehicles in the middle lane. The same principle lies with dual carriageways with more than three lanes.

U.S.-state-specific Practices

In some US states (such as Louisiana, Massachusetts and New York), although there are laws requiring all traffic on a public way to use the right-most lane unless overtaking, this rule is often ignored and seldom enforced on multi-lane roadways. Some states, such as Colorado, use a combination of laws and signs restricting speeds or vehicles on certain lanes to emphasize overtaking only on the left lane, and to avoid a psychological condition commonly called road rage.

In California, cars may use any lane on multi-lane roadways. Drivers moving slower than the general flow of traffic are required to stay in the right-most lanes (by California Vehicle Code (CVC) 21654) to keep the way clear for faster vehicles and thus speed up traffic. However, faster drivers may legally pass in the slower lanes if conditions allow (by CVC 21754). But the CVC also requires trucks to stay in the right lane, or in the right two lanes if the roadway has four or more lanes going in their direction. The oldest freeways in California, and some freeway interchanges, often have ramps on the left, making signs like "TRUCKS OK ON LEFT LANE" or "TRUCKS MAY USE ALL LANES" necessary to override the default rule. Lane splitting, or riding motorcycles in the space between cars in traffic, is permitted as long as it is done in a safe and prudent manner.

One-way Roadways

In order to increase traffic capacity and safety, a route may have two or more separate roads for each direction of traffic. Alternatively, a given road might be declared *one-way*.

High-speed Roads

In large cities, moving from one part of the city to another by means of ordinary streets and avenues can be time-consuming since traffic is often slowed by at-grade junctions, tight turns, narrow marked lanes and lack of a minimum speed limit. Therefore, it has become common practice for larger cities to build roads for faster through traffic. There are two different types of roads used to provide high-speed access across urban areas:

- The controlled-access highway (*freeway* or *motorway*) is a divided multi-lane highway with fully controlled access and grade-separated intersections (no cross traffic). Some freeways are called *expressways*, *super-highways*, or *turnpikes*, depending on local usage. Access to freeways is fully controlled; entering and leaving the freeway is permitted only at grade-separated interchanges.
- The limited-access road (often called *expressway* in areas where the name does not refer to a freeway or motorway) is a lower-grade type of road with some or many of the characteristics of a controlled-access highway: usually a broad multi-lane avenue, frequently divided, with some grade separation at intersections.

Motor vehicle drivers wishing to travel over great distances within the city will usually take the freeways or expressways in order to minimize travel time. When a crossing road is at the same grade as the freeway, a bridge (or, less often, an underpass) will be built for the crossing road. If the freeway is elevated, the crossing road will pass underneath it.

Minimum speed signs are sometimes posted (although increasingly rare) and usually indicate that any vehicle traveling slower than 40 mph (64 km/h) should indicate a slower speed of travel to other motor vehicles by engaging the vehicle's four-way flashing lights. Alternative slower-than-posted speeds may be in effect, based on the posted speed limit of the highway/freeway.

Systems of freeways and expressways are also built to connect distant and regional cities, notable systems include the Interstate highways, the Autobahnen and the Expressway Network of the People's Republic of China.

One-way Streets

In more sophisticated systems such as large cities, this concept is further extended: some streets are marked as being *one-way*, and on those streets all traffic must flow in only one direction. Pedestrians on the sidewalks are generally not limited to one-way movement. Drivers wishing to reach a destination they have already passed must return via other streets. One-way streets, despite the inconveniences to some individual drivers, can greatly improve traffic flow since they usually allow traffic to move faster and tend to simplify intersections.

Congested Traffic

Traffic slows to a crawl on the Monash Freeway in Melbourne, Australia through peak hour traffic.

In some places traffic volume is consistently, extremely large, either during periods of time referred to as *rush hour* or perpetually. Exceptionally, traffic upstream of a vehicular collision or an obstruction, such as construction, may also be constrained, resulting in a traffic jam. Such dynamics in relation to traffic congestion is known as traffic flow. Traffic engineers sometimes gauge the quality of traffic flow in terms of level of service.

In measured traffic data, common spatiotemporal empirical features of traffic congestion have been found that are qualitatively the same for different highways in different countries. Some of these common features distinguish the wide moving jam and synchronized flow phases of congested traffic in Kerner's three-phase traffic theory.

Rush Hour

During business days in most major cities, traffic congestion reaches great intensity at predictable times of the day due to the large number of vehicles using the road at the same time. This phenomenon is called *rush hour* or *peak hour*, although the period of high traffic intensity often exceeds one hour.

Congestion Mitigation

Rush Hour Policies

Some cities adopt policies to reduce rush-hour traffic and pollution and encourage the use of public transportation. For example, in São Paulo, Manila and in Mexico City, each vehicle has a specific day of the week in which it is forbidden from traveling the roads during rush hour. The day for each vehicle is taken from the license plate number, and this rule is enforced by traffic police and also by hundreds of strategically positioned traffic cameras backed by computerized image-recognition systems that issue tickets to offending drivers.

In the United States and Canada, several expressways have a special lane (called an "HOV Lane" – High Occupancy Vehicle Lane) that can only be used by cars carrying two (some locations-three) or more people. Also, many major cities have instituted strict parking prohibitions during rush hour on major arterial streets leading to and from the central business district. During designated weekday hours, vehicles parked on these primary routes are subject to prompt ticketing and towing at owner expense. The purpose of these restrictions is to make available an additional traffic lane in order to maximize available traffic capacity. Additionally, several cities offer a public telephone service where citizens can arrange rides with others depending on where they live and work. The purpose of these policies is to reduce the number of vehicles on the roads and thus reduce rush-hour traffic intensity.

Metered freeways are also a solution for controlling rush hour traffic. In Phoenix, Arizona and Seattle, Washington, among other places, metered on-ramps have been implemented. During rush hour, traffic signals are used with green lights to allow one car per blink of the light to proceed on to the freeway.

Pre-emption

In some areas, emergency responders are provided with specialized equipment, such as a Mobile

Infrared Transmitter, which allows emergency response vehicles, particularly fire-fighting apparatus, to have high-priority travel by having the lights along their route change to green. The technology behind these methods has evolved, from panels at the fire department (which could trigger and control green lights for certain major corridors) to optical systems (which the individual fire apparatus can be equipped with to communicate directly with receivers on the signal head). In other areas, public transport buses have special equipment to get green lights.

During emergencies where evacuation of a heavily populated area is required, local authorities may institute contraflow lane reversal, in which all lanes of a road lead away from a danger zone regardless of their original flow. Aside from emergencies, contraflow may also be used to ease traffic congestion during rush hour or at the end of a sports event (where a large number of cars are leaving the venue at the same time). For example, the six lanes of the Lincoln Tunnel can be changed from three inbound and three outbound to a two/four configuration depending on traffic volume. The Brazilian highways Rodovia dos Imigrantes and Rodovia Anchieta connect São Paulo to the Atlantic coast. Almost all lanes of both highways are usually reversed during weekends to allow for heavy seaside traffic. The reversibility of the highways requires many additional highway ramps and complicated interchanges.

Intelligent Transportation Systems

An intelligent transportation system (ITS) is a system of hardware, software, and operators-in-the-loop that allow better monitoring and control of traffic in order to optimize traffic flow. As the number of vehicle lane miles traveled per year continues to increase dramatically, and as the number of vehicle lane miles constructed per year has not been keeping pace, this has led to ever-increasing traffic congestion. As a cost-effective solution toward optimizing traffic, ITS presents a number of technologies to reduce congestion by monitoring traffic flows through the use of sensors and live cameras or analysing cellular phone data travelling in cars (floating car data) and in turn rerouting traffic as needed through the use of variable message boards (VMS), highway advisory radio, on board or off board navigation devices and other systems through integration of traffic data with navigation systems. Additionally, the roadway network has been increasingly fitted with additional communications and control infrastructure to allow traffic operations personnel to monitor weather conditions, for dispatching maintenance crews to perform snow or ice removal, as well as intelligent systems such as automated bridge de-icing systems which help to prevent accidents.

Traffic Flow

Traffic flow is the theory developed based on flow of vehicle on the lane, and its connections with other vehicles, pedestrians, signals, present on the road. Free movement of traffic is affected by many factors like design speed, percentage of heavy vehicles, number of lanes, and intersections present along the road.

Traffic flow in a road is expressed as number of vehicles using the particular road per unit duration of one hour. It is measured using traffic counts made for a particular duration at one point on the lane stretch. There will be variability in the counts made during different hours of time on a particular day. For any analysis to be made based on traffic flow, peak hour traffic is considered.

Traffic flow measurement is required on the road to locate the point on the road, where congestion happens, assessing the requirement of traffic signal at an intersection, estimating the capacity of the road way to meet with the present flow, etc. Traffic flow is dependent upon two parameters, speed of traffic, and density of vehicles on the road.

For flow to be without congestion, density of vehicles should not be more than 30 vehicles per mile length. Free flow speed on a freeway should be 75 for improved level of service on a particular road. On an urban road, free flow speed should be maintained around 50mph.

Traffic flow is measured using sensors, loop detectors, video measurement, and manual count of vehicles subject to the magnitude of traffic counts. Measured traffic is converted as Annual Average Daily Traffic (AADT), design hourly volume for the purpose of pavement design, provision of lanes, etc.

Traffic Stream Properties

Traffic flow is generally constrained along a one-dimensional pathway (e.g. a travel lane). A time-space diagram shows graphically the flow of vehicles along a pathway over time. Time is displayed along the horizontal axis, and distance is shown along the vertical axis. Traffic flow in a time-space diagram is represented by the individual trajectory lines of individual vehicles. Vehicles following each other along a given travel lane will have parallel trajectories, and trajectories will cross when one vehicle passes another. Time-space diagrams are useful tools for displaying and analyzing the traffic flow characteristics of a given roadway segment over time (e.g. analyzing traffic flow congestion).

There are three main variables to visualize a traffic stream: speed (v), density (indicated k; the number of vehicles per unit of space), and flow (indicated q; the number of vehicles per unit of time).

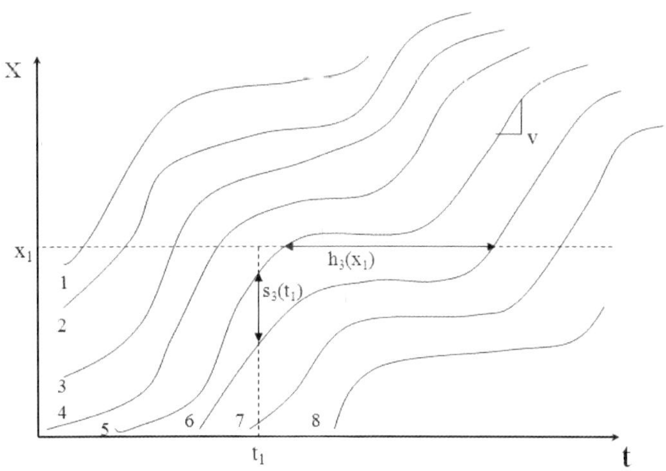

Time Space diagram

Speed

Speed is the distance covered per unit time. One cannot track the speed of every vehicle; so, in practice, average speed is measured by sampling vehicles in a given area over a period of time. Two definitions of average speed are identified: "time mean speed" and "space mean speed".

- "Time mean speed" is measured at a reference point on the roadway over a period of time. In practice, it is measured by the use of loop detectors. Loop detectors, when spread over a reference area, can identify each vehicle and can track its speed. However, average speed measurements obtained from this method are not accurate because instantaneous speeds averaged over several vehicles do not account for the difference in travel time for the vehicles that are traveling at different speeds over the same distance.

$$v_t = (1/m)\sum_{i=1}^{m} v_i$$

where m represents the number of vehicles passing the fixed point and v_i is the speed of the *ith* vehicle.

- "Space mean speed" is measured over the whole roadway segment. Consecutive pictures or video of a roadway segment track the speed of individual vehicles, and then the average speed is calculated. It is considered more accurate than the time mean speed. The data for space calculating space mean speed may be taken from satellite pictures, a camera, or both.

$$v_s = \left((1/n)(\sum_{i=1}^{n}(1/v_i)\right)^{-1}$$

where n represents the number of vehicles passing the roadway segment.

The "space mean speed" is thus the harmonic mean of the speeds.

The time mean speed is never less than space mean speed:

$$v_t = v_s + \frac{\sigma_s^2}{v_s}$$

where σ_s^2 is the variance of the space mean speed

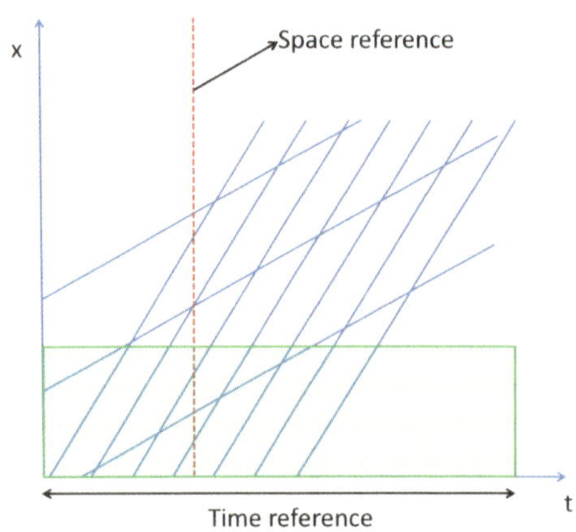

Time Mean Speed vs. Space Mean Speed

Space Mean- and Time Mean speeds

In a time-space diagram, the instantaneous velocity, v = dx/dt, of a vehicle is equal to the slope along the vehicle's trajectory. The average velocity of a vehicle is equal to the slope of the line connecting the trajectory endpoints where a vehicle enters and leaves the roadway segment. The vertical separation (distance) between parallel trajectories is the vehicle spacing (s) between a leading and following vehicle. Similarly, the horizontal separation (time) represents the vehicle headway (h). A time-space diagram is useful for relating headway and spacing to traffic flow and density, respectively.

Density

Density (k) is defined as the number of vehicles per unit length of the roadway. In traffic flow, the two most important densities are the critical density (k_c) and jam density (k_j). The maximum density achievable under free flow is k_c, while k_j is the maximum density achieved under congestion. In general, jam density is seven times the critical density. Inverse of density is spacing (s), which is the center-to-center distance between two vehicles.

$$k = \frac{1}{s}$$

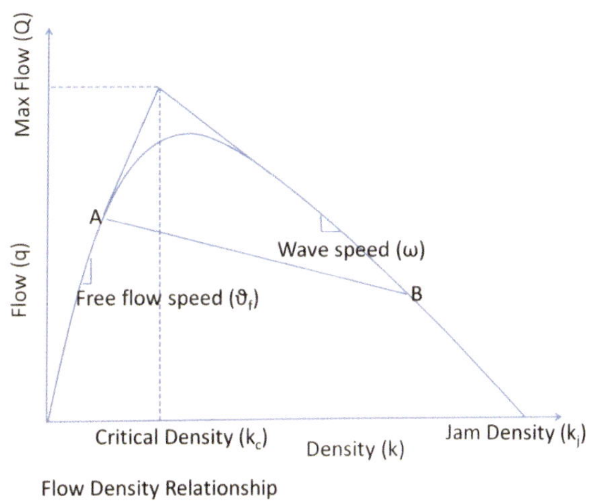

Flow Density relationship

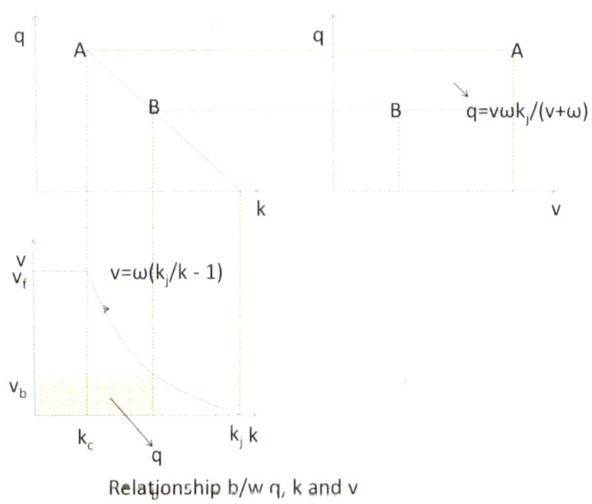

Relationship between flow (q), density (k), and speed (v)

The density (k) within a length of roadway (L) at a given time (t_1) is equal to the inverse of the average spacing of the n vehicles.

$$K(L,t_1)=\frac{n}{L}=\frac{1}{\bar{s}(t_1)}$$

In a time-space diagram, the density may be evaluated in the region A.

$$k(A)=\frac{n}{L}=\frac{ndt}{Ldt}=\frac{tt}{|A|}$$

where tt is the total travel time in A.

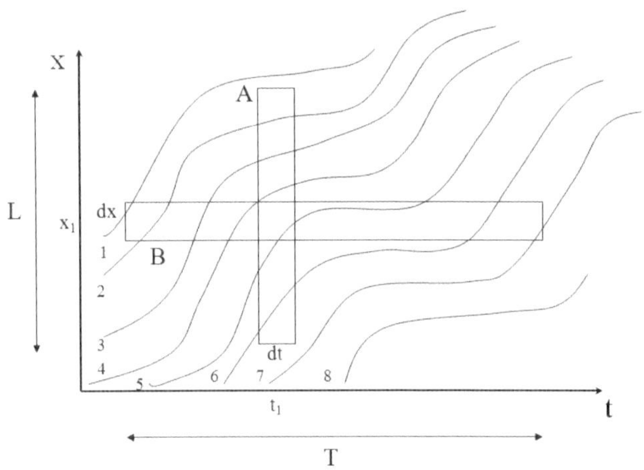

Figure

Flow

Flow (q) is the number of vehicles passing a reference point per unit of time, vehicles per hour. The inverse of flow is headway (h), which is the time that elapses between the ith vehicle passing a reference point in space and the (i + 1)th vehicle. In congestion, h remains constant. As a traffic jam forms, h approaches infinity.

$$q=kv$$

$$q=1/h$$

The flow (q) passing a fixed point (x_1) during an interval (T) is equal to the inverse of the average headway of the m vehicles.

$$q(T,x_1)=\frac{m}{T}=\frac{1}{\bar{h}(x_1)}$$

In a time-space diagram, the flow may be evaluated in the region B.

$$q(B)=\frac{m}{T}=\frac{mdx}{Tdx}=\frac{td}{|B|}$$

where *td* is the total distance traveled in *B*.

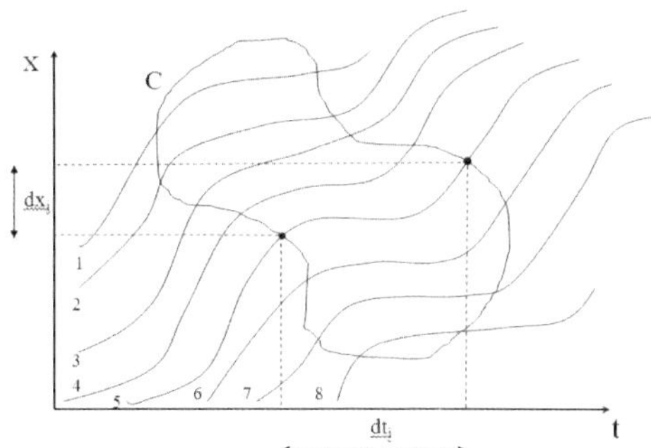

Figure

Generalized Density and Flow in Time-space Diagram

A more general definition of the flow and density in a time-space diagram is illustrated by region C:

$$q(C) = \frac{td}{|C|}$$

$$k(C) = \frac{tt}{|C|}$$

where:

$$td = \sum_{i=1}^{m} dx_i$$

$$tt = \sum_{i=1}^{n} dt_i$$

Congestion Shockwave

In addition to providing information on the speed, flow, and density of traffic streams, time-space diagrams may illustrate the propagation of congestion upstream from a traffic bottleneck (shockwave). Congestion shockwaves will vary in propagation length, depending upon the upstream traffic flow and density. However, shockwaves will generally travel upstream at a rate of approximately 20 km/h.

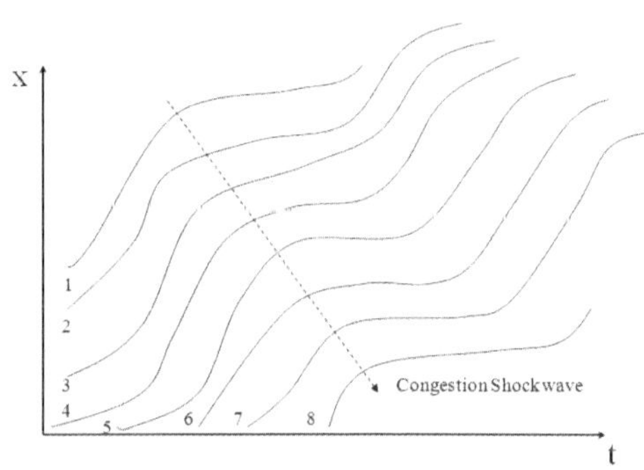

Figure

Stationary Traffic

Traffic on a stretch of road is said to be stationary if an observer does not detect movement in an arbitrary area of the time-space diagram. Traffic is stationary if all the vehicle trajectories are parallel and equidistant. It is also stationary if it is a superposition of families of trajectories with these properties (e.g. fast and slow drivers). By using a very small hole in the template one could sometimes view an empty region of the diagram and other times not, so that even in these cases, one could say that traffic was not stationary. Clearly, for such fine level of observation, stationary traffic does not exist. A microscopic level of observation must be excluded from the definition if traffic appears to be similar through larger windows. In fact, we relax the definition even further by only requiring that the quantities t(A) and d(A) be approximately the same, regardless of where the "large" window (A) is placed.

Methods of Analysis

Analysts approach the problem in three main ways, corresponding to the three main scales of observation in physics:

- Microscopic scale: At the most basic level, every vehicle is considered as an individual. An equation can be written for each, usually an ordinary differential equation (ODE). Cellular automation models can also be used, where the road is divided into cells, each of which contains a moving car, or is empty. The Nagel–Schreckenberg model is a simple example of such a model. As the cars interact it can model collective phenomena such as traffic jams.
- Macroscopic scale: Similar to models of fluid dynamics, it is considered useful to employ a system of partial differential equations, which balance laws for some gross quantities of interest; e.g., the density of vehicles or their mean velocity.
- Mesoscopic (kinetic) scale: A third, intermediate possibility, is to define a function $f(t,x,V)$ which expresses the probability of having a vehicle at time t in position x which runs with velocity V. This function, following methods of statistical mechanics, can be computed using an integro-differential equation such as the Boltzmann equation.

The engineering approach to analysis of highway traffic flow problems is primarily based on empirical analysis (i.e., observation and mathematical curve fitting). One major reference used by American planners is the *Highway Capacity Manual*, published by the Transportation Research Board, which is part of the United States National Academy of Sciences. This recommends modelling traffic flows using the whole travel time across a link using a delay/flow function, including the effects of queuing. This technique is used in many US traffic models and in the SATURN model in Europe.

In many parts of Europe, a hybrid empirical approach to traffic design is used, combining macro-, micro-, and mesoscopic features. Rather than simulating a steady state of flow for a journey, transient "demand peaks" of congestion are simulated. These are modeled by using small "time slices" across the network throughout the working day or weekend. Typically, the origins and destinations for trips are first estimated and a traffic model is generated before being calibrated by comparing the mathematical model with observed counts of actual traffic flows, classified by type of vehicle. "Matrix estimation" is then applied to the model to achieve a better match to observed link counts before any changes, and the revised model is used to generate a more realistic traffic forecast for any proposed scheme. The model would be run several times (including a current baseline, an "average day" forecast based on a range of economic parameters and supported by sensitivity analysis) in order to understand the implications of temporary blockages or incidents around the network. From the models, it is possible to total the time taken for all drivers of different types of vehicle on the network and thus deduce average fuel consumption and emissions.

Much of UK, Scandinavian, and Dutch authority practice is to use the modelling program CONTRAM for large schemes, which has been developed over several decades under the auspices of the UK's Transport Research Laboratory, and more recently with the support of the Swedish Road Administration. By modelling forecasts of the road network for several decades into the future, the economic benefits of changes to the road network can be calculated, using estimates for value of time and other parameters. The output of these models can then be fed into a cost-benefit analysis program.

Cumulative Vehicle Count Curves (N-curves)

A cumulative vehicle count curve, the *N*-curve, shows the cumulative number of vehicles that pass a certain location x by time t, measured from the passage of some reference vehicle. This curve can be plotted if the arrival times are known for individual vehicles approaching a location x, and the departure times are also known as they leave location x. Obtaining these arrival and departure times could involve data collection: for example, one could set two point sensors at locations X_1 and X_2, and count the number of vehicles that pass this segment while also recording the time each vehicle arrives at X_1 and departs from X_2. The resulting plot is a pair of cumulative curves where the vertical axis (N) represents the cumulative number of vehicles that pass the two points: X_1 and X_2, and the horizontal axis (t) represents the elapsed time from X_1 and X_2.

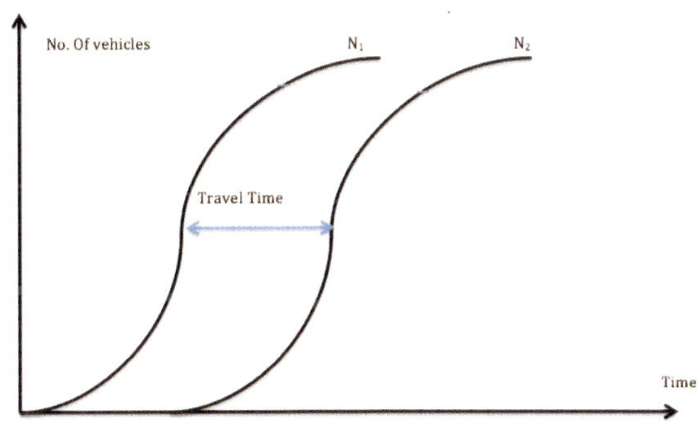

Figure: Simple cumulative curves

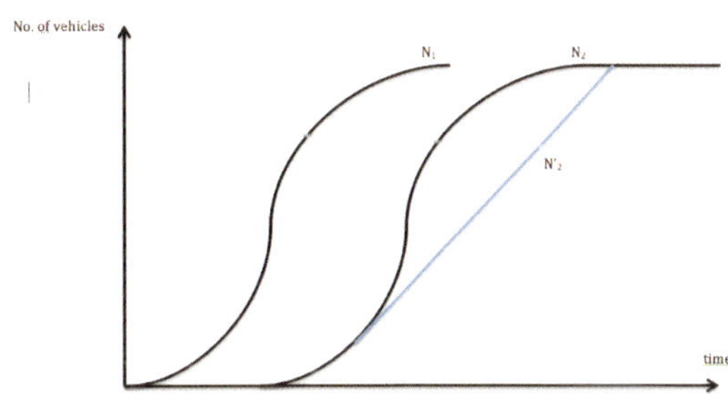

Figure: Arrival, virtual arrival, and departure curves

If vehicles experience no delay as they travel from X_1 to X_2, then the arrivals of vehicles at location X_1 is represented by curve N_1 and the arrivals of the vehicles at location X_2 is represented by N_2 in figure. More commonly, curve N_1 is known as the *arrival curve* of vehicles at location X_1 and curve N_2 is known as the *arrival curve* of vehicles at location X_2. Using a one-lane signalized approach to an intersection as an example, where X_1 is the location of the stop bar at the approach and X_2 is an arbitrary line on the receiving lane just across of the intersection, when the traffic signal is green, vehicles can travel through both points with no delay and the time it takes to travel that distance is equal to the free-flow travel time. Graphically, this is shown as the two separate curves in figure.

However, when the traffic signal is red, vehicles arrive at the stop bar (X_1) and are delayed by the red light before crossing X_2 some time after the signal turns green. As a result, a queue builds at the stop bar as more vehicles are arriving at the intersection while the traffic signal is still red. Therefore, for as long as vehicles arriving at the intersection are still hindered by the queue, the curve N_2 no longer represents the vehicles' arrival at location X_2; it now represents the vehicles' *virtual arrival* at location X_2, or in other words, it represents the vehicles' arrival at X_2 if they did not experience any delay. The vehicles' arrival at location X_2, taking into account the delay from the traffic signal, is now represented by the curve N'_2 in figure.

However, the concept of the *virtual arrival curve* is flawed. This curve does not correctly show the queue length resulting from the interruption in traffic (i.e. red signal). It assumes that all vehicles are still reaching the stop bar before being delayed by the red light. In other words, the *virtual arrival curve* portrays the stacking of vehicles vertically at the stop bar. When the traffic signal turns green, these vehicles are served in a first-in-first-out (FIFO) order. For a multi-lane approach, however, the service order is not necessarily FIFO. Nonetheless, the interpretation is still useful because of the concern with average total delay instead of total delays for individual vehicles.

Step Function vs. Smooth Function

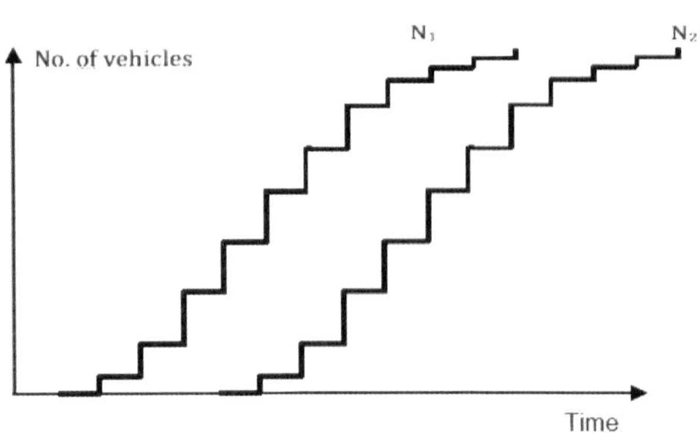

Figure: Step function

The traffic light example depicts *N*-curves as smooth functions. Theoretically, however, plotting *N*-curves from collected data should result in a step-function. Each step represents the arrival or departure of one vehicle at that point in time. When the *N*-curve is drawn on larger scale reflecting a period of time that covers several cycles, then the steps for individual vehicles can be ignored, and the curve will then look like a smooth function.

N-curve: Traffic Flow Characteristics

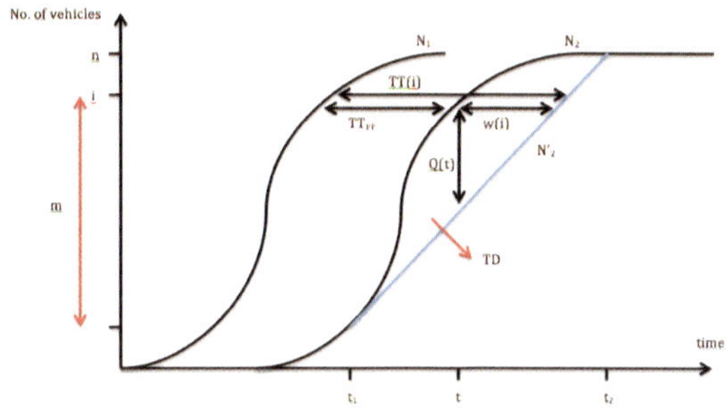

Figure: Traffic flow characteristics from two *N*-curves

The *N*-curve can be used in a number of different traffic analyses, including freeway bottlenecks and dynamic traffic assignment. This is due to the fact that a number of traffic flow characteristics can be derived from the plot of cumulative vehicle count curves. Illustrated in figure below are the different traffic flow characteristics that can be derived from the *N*-curves.

These are the different traffic flow characteristics from figure:

Symbol	Definition
N_1	the cumulative number of vehicles arriving at location X_1
N_2	the virtual cumulative number of vehicles arriving at location X_2, or the cumulative number of vehicles that would have liked to cross X_2 by time t
N'_2	the actual cumulative number of vehicles arriving at location X_2
TT_{FF}	the time it takes to travel from location X_1 to location X_2 at free-flow conditions
$w(i)$	the delay experienced by vehicle i as it travels from X_1 to X_2
$TT(i)$	the total time it takes to travel from X_1 to X_2 including delays ($TT_{FF} + w(i)$)
$Q(t)$	the queue at any time t, or the number of vehicles being delayed at time t
n	total number of vehicles in the system
m	total number of delayed vehicles
TD	total delay experienced by m vehicles (area between N_2 and N'_2)
t_1	time at which congestion begins
t_2	time at which congestion ends

From these variables, the average delay experienced by each vehicle and the average queue length at any time t can be calculated, using the following formulas:

$$average\ delay\ (w_{avg}) = \frac{total\ delay\ experienced\ by\ m\ vehicles}{total\ number\ of\ delayed\ vehicles} = \frac{TD}{m}$$

$$average\ queue\ (Q_{avg}) = \frac{total\ delay\ experienced\ by\ m\ vehicles}{duration\ of\ congestion} = \frac{TD}{(t_2 - t_1)}$$

Hamilton Jacobi PDE

In traffic flow area, an alternative way to solve the kinematic wave model is to treat it like a Hamilton–Jacobi equation, which is particularly useful in identifying conserved quantities for mechanical systems.

Suppose we are interested in finding the cumulative curve as a function of time and space, *N(t,x)*. Based the definition of cumulative curve, $\frac{\partial N(t,x)}{\partial t} = \rho(x,t)$ refers to the flow and $\frac{\partial N(t,x)}{\partial x} = -k(x,t)$ refers to the density. Note that the sign convention should be consistent. Then the fundamental flow-density ($q-k$) equation: $q = F(k)$ can be expressed in

cumulative count form as:

$$\frac{\partial N(t,x)}{\partial t} - F(-\frac{\partial N(t,x)}{\partial x}) = 0$$

$s.t. N(B) = G(B)$, where B is a known boundary.

Now, for a generic random point $P(x,t)$ in the time-space diagram, the solution to the above partial derivative equation is equivalent to solve the following optimization problem, which minimizes the vehicles pass-by: $N(P) = \min G(x_b) + (t-t_b)R(\frac{x-x_b}{t-t_b})$, where b is a random point on the boundary B.

The function R is defined as the maximum passing rate along the observers. In the case of triangular fundamental diagram, we have $R(v) = Q - K_c * v$. The observer speed $v \in [-w,u]$. Here, notation Q corresponds to the capacity, k_c corresponds to the critical density, u and w are free flow speed and wave speed respectively.

With that being said, the minimization function above simplifies into:

$N(P) = \min G(x_b) + (t-t_b)Q - (x-x_b)K$, where b is a random point on the boundary B. Here, we limit the solution discussion over initial value problems (IVP) and boundary value problems (BVP).

Initial Value Problem

Initial value problem occurs when boundary condition is given at a fixed time, e.g. at $t = 0$ and boundary $B := N(0,x)$. As the observer sp eed is bounded by $v \in [-w,u]$, the potential solution is delimited by two lines $x = x_U + ut$ and $x = x_D - wt$.

Thus, the IVP is defined as follow:

$$N(P) = \min f(x_b) \equiv G(x_b) + (t-t_b)Q - (x-x_b)K_c$$

$$s.t. N(0,x) = G(x)$$

$$x_U < x_b < x_D$$

The local minimum point occurs when the first order derivative is 0 and the second order derivative is greater than 0. Or, the minimum happens at boundaries. So, the set of potential solutions goes as follow:

1. $\forall x_b$ that $f'(x_b) = G'(x_b) + K_c = -k(0, x_b) + K_c = 0$ and $f''(x_b) = G''(x_b) = -k_x(0, x_b) > 0$
2. x_U and x_D.

The solution will be the minimum corresponding $N(P)$ of all candidate points.

$N(P) = \min(f(x_U), f(x_D)$, and all $f(x_b)$ from condition 1).

Specifically, if the initial condition $G(x)$ is a linear function, $N(P) = \min(f(x_U), f(x_D))$

Boundary Value Problem

Similarly, the boundary value problem indicates the boundary condition is given at a fix location, e.g. $B := N(x_0, 0)$. Still, the observer speed is bounded by $v \in [-w, u]$. For a random point $P(x,t)$, the upper bound for solution candidates: if $x > x_0$, $t_U = t - \frac{x - x_0}{u}$; else, $t_D = t - \frac{x_0 - x}{w}$.

The BVP is defined as follow:

$$N(P) = \min f(t_b) \equiv G(t_b) + (t - t_b)Q - (x - x_b)K_c$$

$$s.t. N(t, x_0) = G(t)$$

$$\begin{cases} t < t_U, & if\, x > x_0 \\ t < t_D, & if\, x < x_0 \end{cases}$$

The first order derivative: $f'(t_b) = G'(t_b) - Q = q(t_b, 0) - Q$ is always smaller than 0 because flows won't exceed the capacity. Thus, the minimum happens at the upper bound of the time axis.

$$N(P) = \begin{cases} f(t_U), & if\, x > x_0 \\ f(t_D), & if\, x < x_0 \end{cases}$$

In practice, people use this method to estimate the traffic states at $P(t,x)$ between two loop detectors, which can be viewed as a combination of two boundary value problems (one at upstream and one at down stream). Denote the upstream loop detector location as and the downstream loop detector location as . Based on the conclusion above, the minimum value occurs at the upper bound along the time axis.

$$N(P) = \min(f(t_U), f(t_D)), \text{ with } \begin{cases} t_U = t - \frac{x - x_U}{u} \\ t_D = t - \frac{x_D - x}{w} \end{cases}$$

Applications

The Bottleneck Model

One application of the *N*-curve is the bottleneck model, where the cumulative vehicle count is known at a point *before* the bottleneck (i.e. this is location X_1). However, the cumulative vehicle count is not known at a point *after* the bottleneck (i.e. this is location X_2), but rather only the capacity of the bottleneck, or the discharge rate, μ, is known. The bottleneck model can be applied to real-world bottleneck situations such as those resulting from a roadway design problem or a traffic incident.

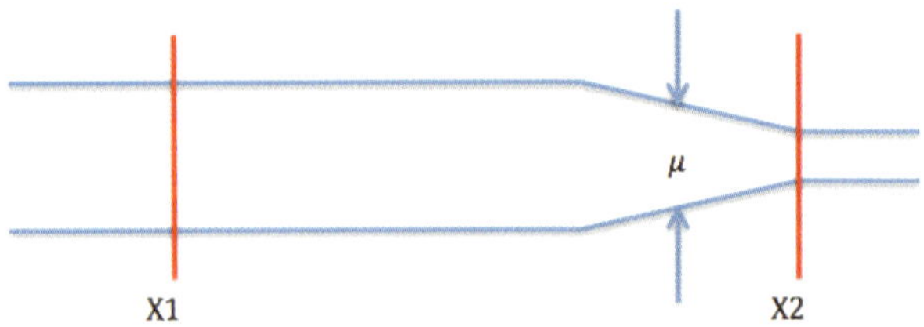

Figure: Roadway section experiencing a bottleneck

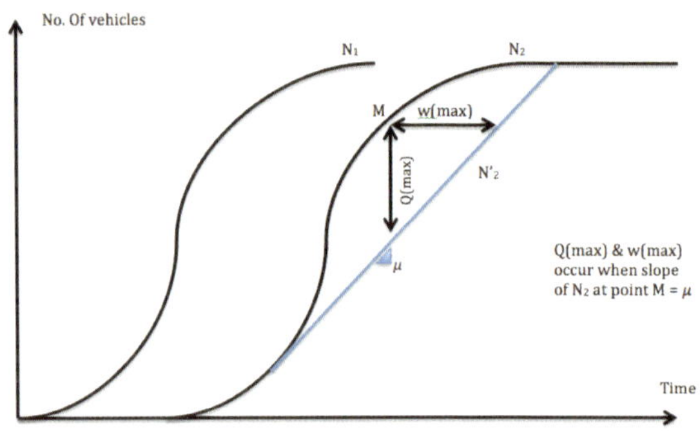

Figure: Maximum queue length and delay

Take a roadway section where a bottleneck exists such as in figure. At some location X_1 before the bottleneck, the arrivals of vehicles follow a regular *N*-curve. If the bottleneck is absent, the departure rate of vehicles at location X_2 is essentially the same as the arrival rate at X_1 at some later time (i.e. at time TT_{FF} – free-flow travel time). However, due to the bottleneck, the system at location X_2 is now only able to have a departure rate of μ. When graphing this scenario, essentially we have the same situation as in figure, where the arrival curve of vehicles is N_1, the departure curve of vehicles absent the bottleneck is N_2, and the limited departure curve of vehicles given the bottleneck is N'_2. The discharge rate μ is the slope of curve N'_2, and all the same traffic flow characteristics as in figure can be determined from this diagram. The maximum delay and maximum queue length can be found at a point *M* in figure where the slope of N_2 is the same as the slope of N'_2; i.e. when the virtual arrival rate is equal to the discharge / departure rate μ.

The *N*-curve in the bottleneck model may also be used to calculate the benefits in removing the bottleneck, whether in terms of a capacity improvement or removing an incident to the side of the roadway.

Tandem Queues

As introduced in the section above, the N-curve is an applicable model to estimate traffic delay during time by setting arrival and departure cumulative counting curve. Since the curve can represent various traffic characteristics and roadway conditions, the delay and queue situations under these conditions will be able to be recognized and modeled using N-curves. Tandem queues occur when multiple bottlenecks exist between the arrival and departure locations. Figure shows a qualitative layout of a tandem-queue roadway segment with a certain initial arrival. The bottlenecks along the stream have their own capacity, 'μ_i [veh/time], and the departure is defined at the downstream end of the entire segment.

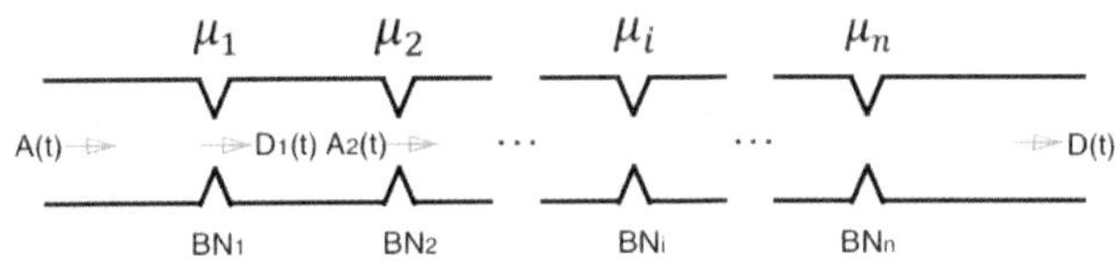

Figure: Tandem Queues

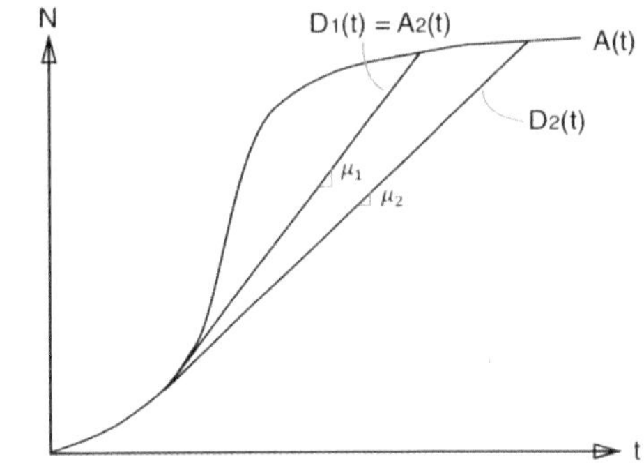

Figure: N-Curve of Tandem Queues with Two BNs

To determine the ultimate departure, $D(t)$, it can be an available method to research on the individual departures, $D_i(t)$. As shown in the Figure, if the free-flow travel-time is neglected, the departure of BN_{i-1} will be the virtual arrival of BN_i, which can also be presented as $D_{i-1}(t) = A_i(t)$. Thus, the N-curve of a roadway with two bottlenecks (minimum number of BNs along a tandem-queue roadway) can be developed as Figure with $\mu_1 < \mu_2$. In this case, $D_2(t)$ will be the ultimate departure of this 2-BN tandem-queue roadway.

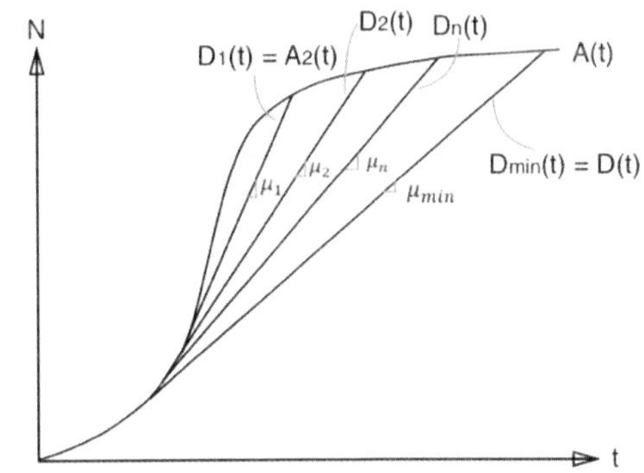

Figure: N-Curve of Tandem Queues with n BNs

Regarding of a tandem-queue roadway having 3 BNs with $\mu_1 < \mu_2$, *if* $\mu_1 < \mu_2 < \mu_3$, similarly as the 2-BN case, $D_3(t)$ will be the ultimate departure of this 3-BN tandem-queue roadway. If, however, $\mu_1 < \mu_2 < \mu_3$, $D_2(t)$ will then still be the ultimate departure of the 3-BN tandem-queue roadway. Thus, it can be summarized that, the departure of the bottleneck with the minimum capacity will be the ultimate departure of the entire system, regardless of the other capacities and the number of bottlenecks. Figure above shows a general case with n BNs.

The N-curve model describing above represents a significant characteristic of the tandem-queue systems, which is that the ultimate departure only depends on the bottleneck with the minimum capacity. In a practical perspective, when the resources (economy, effort, etc.) of the investment on tandem-queue systems are limited, the investment can mainly focus on the bottleneck with the worst condition.

Traffic Light

A signalized intersection will have special departure behaviors. With simplified speaking, a constant releasing free-flow capacity, μ_s, exists during the green phases. On the contrary, the releasing capacity during the red phases should be zero. Thus, the departure N-curve regardless of arrival will look like as Figure below: counts increase with the slope of μ_s during green, and remain the same during red..

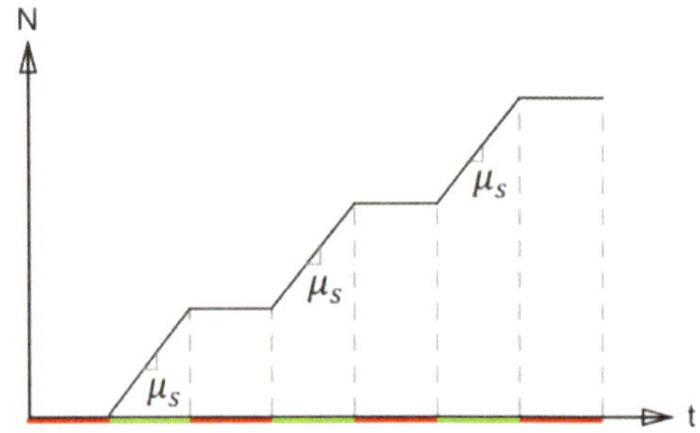

Figure: Departure curve for a signal with a releasing capacity

Saturated case of a traffic light occurs when the releasing capacity is fully used. This case usually exists when the arriving demand is relatively large. The N-curve representation of the saturated case is shown in the Figure below.

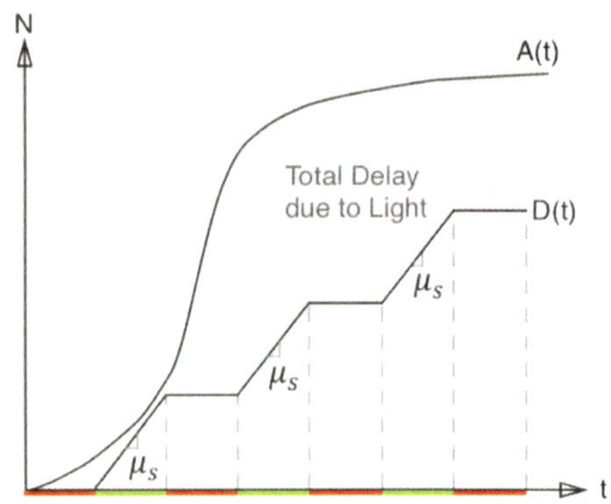

Figure: Saturated case at a traffic light

Unsaturated case of a traffic light occurs when releasing capacity is not fully used. This case usually exists when the arriving demand is relatively small. The N-curve representation of the unsaturated case is shown in the Figure below. If there is a bottleneck with a capacity of $\mu_b\,(<\mu_s)$ downstream of the light, the ultimate departure of the light-bottleneck system will be that of the downstream bottleneck.

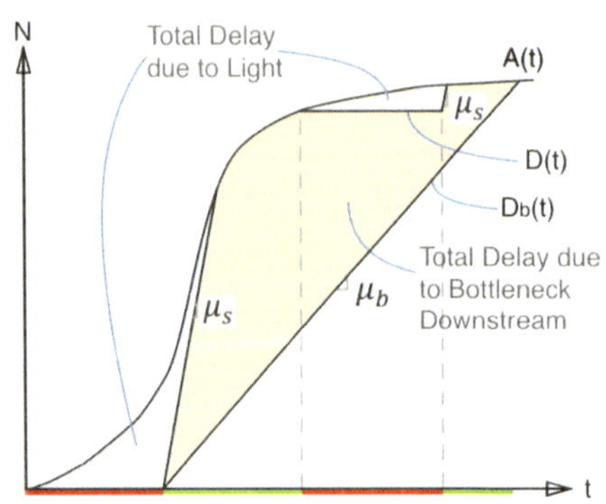

Figure: Unsaturated case at a traffic light with a downstream bottleneck

Dynamic Traffic Assignment

Dynamic traffic assignment can also be solved using the *N*-curve. There are two main approaches to tackle this problem: system optimum, and user equilibrium. This application will be discussed further in the following section.

Traffic Assignment

The aim of traffic flow analysis is to create and implement a model which would enable vehicles to reach their destination in the shortest possible time using the maximum roadway capacity. This is a four-step process:

- Generation – the program estimates how many trips would be generated. For this, the program needs the statistical data of residence areas by population, location of workplaces etc.;
- Distribution – after generation it makes the different Origin-Destination (OD) pairs between the location found in step 1;
- Modal Split/Mode Choice – the system has to decide how much percentage of the population would be split between the difference modes of available transport, e.g. cars, buses, rails, etc.;
- Route Assignment – finally, routes are assigned to the vehicles based on minimum criterion rules.

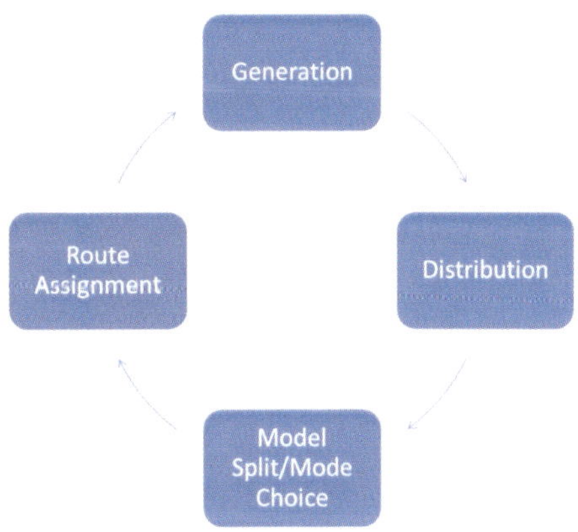

4 Step Model for Traffic Assignment

Figure: The Four Step Travel Demand Model for Traffic Assignment

This cycle is repeated until the solution converges.

There are two main approaches to tackle this problem with the end objectives:

- System optimum
- User equilibrium

System Optimum

System Optimum is based on the assumption that routes of all vehicles would be controlled by the system, and that rerouting would be based on maximum utilization of resources and minimum total system cost. (Cost can be interpreted as travel time.) Hence, in a System Optimum routing algorithm, all routes between a given OD pair have the same marginal cost. In traditional transportation economics, System Optimum is determined by equilibrium of demand function and marginal cost function. In this approach, marginal cost is roughly depicted as increasing function in traffic congestion. In traffic flow approach, the marginal cost of the trip can be expressed as sum of the cost(delay time, w) experienced by the driver and the externality(e) that a driver imposes on the rest of the users. Suppose there is a freeway(0) and an alternative route(1), which users can be diverted onto off-ramp. Operator knows total arrival rate(A(t)), the capacity of the freeway(μ_0), and the capacity of the alternative route(μ_1). From the time 't_0', when freeway is congested, some of the users start moving to alternative route. However, when 't_1', alternative route is also full of capacity. Now operator decides the number of vehicles(N), which use alternative route. The optimal number of vehicles(N) can be obtained by calculus of variation, to make marginal cost of each route equal. Thus, optimal condition is $T_0 = T_1 + \Delta_1$. In this graph, we can see that the queue on the alternative route should clear Δ_1 time units before it clears from the freeway. This solution does not define how we should allocates vehicles arriving between t_1 and T_1, we just can conclude that the optimal solution is not unique. If operator wants freeway not to be congested, operator can impose the congestion toll, $e_0 - e_1$, which is the difference between the externality of freeway and alternative route. In this situation, freeway will maintain free flow speed, however alternative route will be extremely congested.

User Equilibrium

The user optimum equilibrium assumes that all users choose their own route towards their destination based on the travel time that will be consumed in different route options. The users will choose the route which requires the least travel time. The user optimum model is often used in simulating the impact on traffic assignment by highway bottlenecks. When the congestion occurs on highway, it will extend the delay time in travelling through the highway and create a longer travel time. Under the user optimum assumption, the users would choose to wait until the travel time using a certain freeway is equal to the travel time using city streets, and hence equilibrium is reached. This equilibrium is called User Equilibrium, Wardrop Equilibrium or Nash Equilibrium.

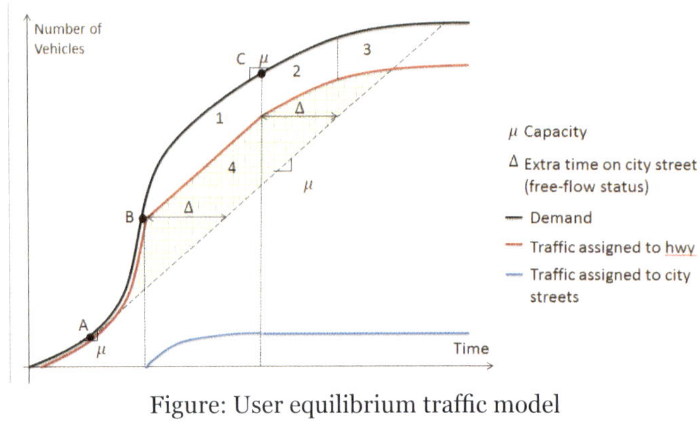

Figure: User equilibrium traffic model

The core principle of User Equilibrium is that all used routes between a given OD pair have the same travel time. An alternative route option is enabled to use when the actual travel time in the system has reached the free-flow travel time on that route.

For a highway user optimum model considering one alternative route, a typical process of traffic assignment is shown in figure above. When the traffic demand stays below the highway capacity, the delay time on highway stays zero. When the traffic demand exceeds the capacity, the queue of vehicle will appear on the highway and the delay time will increase. Some of users will turn to the city streets when the delay time reaches the difference between the free-flow travel time on highway and the free-flow travel time on city streets. It indicates that the users staying on the highway will spend as much travel time as the ones who turn to the city streets. At this stage, the travel time on both the highway and the alternative route stays the same. This situation may be ended when the demand falls below the road capacity, that is the travel time on highway begins to decrease and all the users will stay on the highway. The total of part area 1 and 3 represents the benefits by providing an alternative route. The total of area 4 and area 2 shows the total delay cost in the system, in which area 4 is the total delay occurs on the highway and area 2 is the extra delay by shifting traffic to city streets.

Time Delay

Both User Optimum and System Optimum can be subdivided into two categories on the basis of the approach of time delay taken for their solution:

- Predictive Time Delay
- Reactive Time Delay

Predictive time delay is based on the concept that the system or the user knows when the congestion point is reached or when the delay of the freeway would be equal to the delay on city streets, and the decision for route assignment is taken in time. On the other hand, reactive time delay is when the system or user waits to experience the point where the delay is observed and the diversion of routes is in reaction to that experience. Predictive delay gives significantly better results than the reactive delay method.

Kerner's Network Breakdown Minimization (BM) Principle

Kerner introduced an alternative approach to traffic assignment based on his network breakdown minimization (BM) principle. Rather than an explicit minimization of travel time that is the objective of System Optimum and User Equilibrium, the BM principle minimizes the probability of the occurrence of congestion in a traffic network. Under sufficient traffic demand, the application of the BM principle should lead to implicit minimization of travel time in the network

Traffic Bottleneck

Traffic bottlenecks are disruptions of traffic on a roadway caused either due to road design, traffic lights, or accidents. There are two general types of bottlenecks, stationary and moving bottlenecks. Stationary bottlenecks are those that arise due to a disturbance that occurs due to a stationary situation like narrowing of a roadway, an accident. Moving bottlenecks on the other hand are those vehicles or vehicle behavior that causes the disruption in the vehicles which are upstream of the vehicle. Generally moving bottlenecks are caused by heavy trucks as they are slow moving vehicles with less acceleration and also may make lane changes.

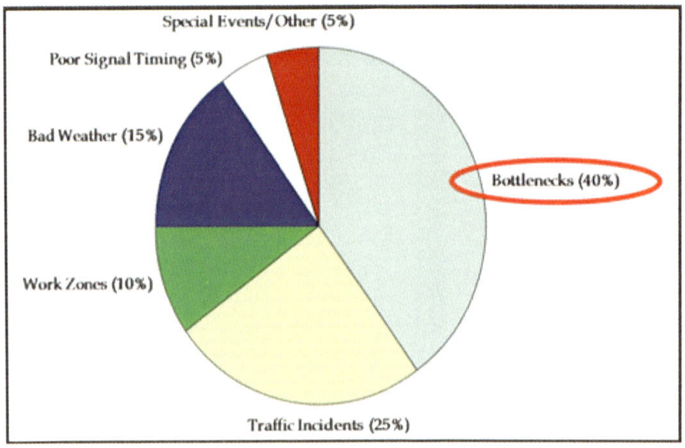

Figure

Bottlenecks are important considerations because they impact the flow in traffic, the average speeds of the vehicles. The main consequence of a bottleneck is an immediate reduction in capacity of the roadway. The Federal Highway Authority has stated that 40% of all congestion is from bottlenecks figure above shows the pie-chart for various causes of congestion.

Stationary Bottleneck

The general cause of stationary bottlenecks are lane drops which occurs when the a multilane roadway loses one or more its lane. This causes the vehicular traffic in the ending lanes to merge onto the other lanes.

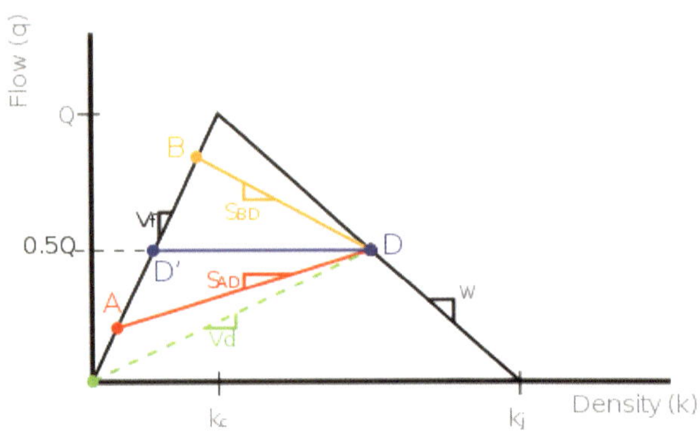

Figure

Consider a stretch of highway with two lanes in one direction. Suppose that the fundamental diagram is modeled as shown here. The highway has a peak capacity of Q vehicles per hour, corresponding to a density of k_c vehicles per mile. The highway normally becomes jammed at k_j vehicles per mile.

Before capacity is reached, traffic may flow at A vehicles per hour, or a higher B vehicles per hour. In either case, the speed of vehicles is v_f, or "free flow," because the roadway is under capacity.

Now, suppose that at a certain location x_0, the highway narrows to one lane. The maximum capacity is now limited to D', or half of Q, since only one lane of the two is available. D shares the same flowrate as state D', but its vehicular density is higher.

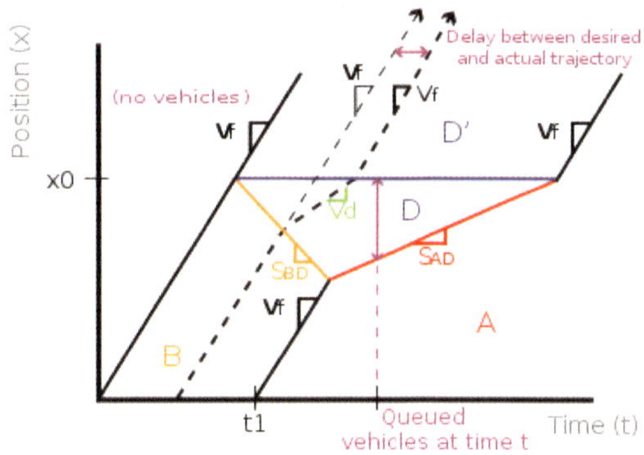

Figure

Using a time-space diagram, we may model the bottleneck event. Suppose that at time 0, traffic begins to flow at rate B and speed v_f. After time t1, vehicles arrive at the lower flowrate A.

Before the first vehicles reach location x_0, the traffic flow is unimpeded. However, downstream of x_0, the roadway narrows, reducing the capacity by half – and to below that of state B. Due to this, vehicles will begin queuing upstream of x_0. This is represented by high-density state D. The vehicle speed in this state is the slower v_d, as taken from the fundamental diagram. Downstream of the bottleneck, vehicles transition to state D', where they again travel at free-flow speed v_f.

Once vehicles arrive at rate A starting at t1, the queue will begin to clear and eventually dissipate. State A has a flowrate below the one-lane capacity of states D and D'.

On the time-space diagram, a sample vehicle trajectory is represented with a dotted arrow line. The diagram can readily represent vehicular delay and queue length. It is a simple matter of taking horizontal and vertical measurements within the region of state D.

Moving Bottleneck

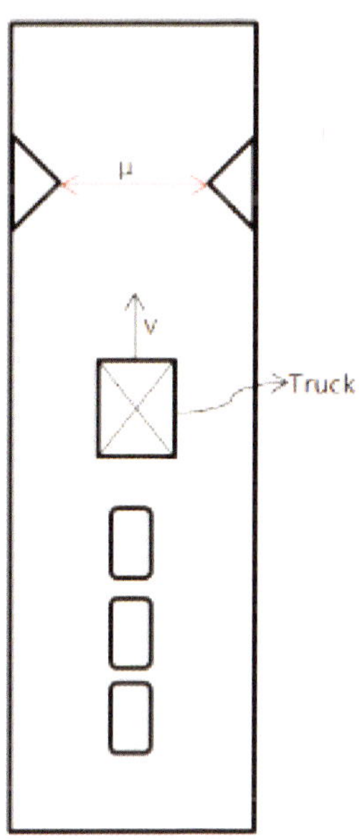

Figure

As explained above, moving bottlenecks are caused due to slow moving vehicles that cause disruption in traffic. Moving bottlenecks can be active or inactive bottlenecks. If the reduced capacity(q_u) caused due to a moving bottleneck is greater than the actual capacity(μ) downstream of the vehicle, then this bottleneck is said to be an active bottleneck. Figure below shows the case of a truck moving with velocity 'v' approaching an downstream location with capacity 'μ'. If the reduced capacity of the truck (q_u) is less than the downstream capacity, then the truck becomes an inactive bottleneck.

Laval 2009, presents a framework for estimating analytical expressions for the capacity reductions caused by a subset of vehicles forced to slow down at horizontal/vertical curves on multilane freeway. In each of the lane the underperforming stream is described in terms of its desired speed distribution and is modeled as per Newell's kinematic wave theory for moving bottlenecks. Lane changing in the presence of trucks can lead to a positive or negative impact on capacity. If the target lane is empty then the lane-changing increases capacity

Figure: A slow tractor creates a moving bottleneck.

For this example, consider three lanes of traffic in one direction. Assume that a truck starts traveling at speed v, slower than the free flow speed v_f. As shown on the fundamental diagram below, q_u represents the reduced capacity (2/3 of Q, or 2 of 3 lanes available) around the truck.

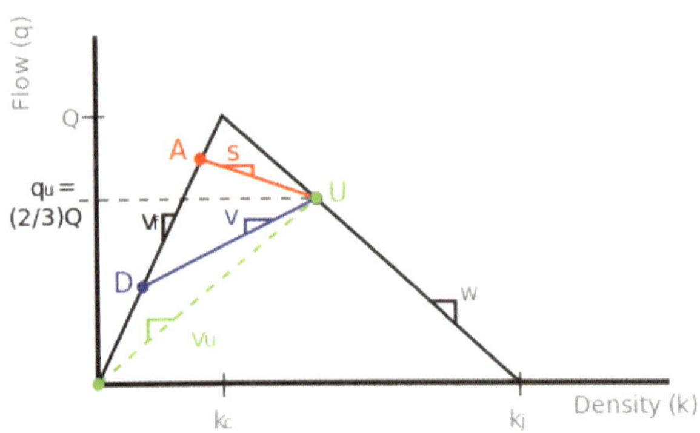

Figure

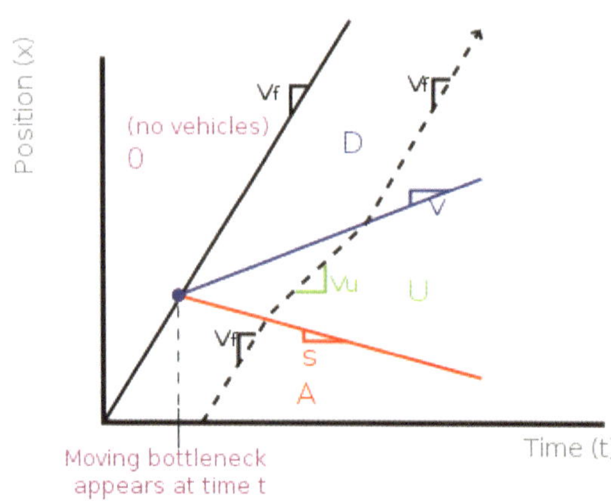

Figure

State A represents normal approaching traffic flow, again at speed v_f. State U, with flowrate q_u, corresponds to the queuing upstream of the truck. On the fundamental diagram, vehicle speed v_u is slower than v_f. But once drivers have navigated around the truck, they can again speed up and transition to downstream state D. While this state travels at free flow, the vehicle density is less because fewer vehicles get around the bottleneck.

Suppose that, at time t, the truck slows from free-flow to v. A queue builds behind the truck, represented by state U. Within the region of state U, vehicles drive slower as indicated by the sample trajectory. Because state U limits to a smaller flow than state A, the queue will back up behind the truck and eventually crowd out the entire highway (slope s is negative). If state U had the higher flow, there would still be a growing queue. However, it would not back up because the slope s would be positive.

Riemann's Problem

Imagine a scenario in which a two lane road is reduced to one lane at point x_0 from here on the road's capacity is reduced to half its original (½μ), Case I. Later along the road at point x_1 the 2nd lane is opened and the capacity is restored to its original (μ), Case II.

Case I

There is a bottleneck limiting the flow of traffic which causes an increase in the density of cars (k) at location (x_0). This causes a deceleration for all oncoming cars traveling at speed u to slow to speed v_d. This shockwave will travel at the speed of the slope of line U-D on the fundamental diagram. The wave speed can be calculated as $v_{shock} = (q_D - q_U)/(k_D - k_U)$. This line delineates the congestion traffic from oncoming free-flow traffic. If the slope of U-D on the fundamental diagram is positive congestion will continue downstream of the highway. If it has a negative slope the congestion will continue upstream. This deceleration is the case I of Riemann's problem.

Case II

In case II of Riemann's problem traffic goes from congestion to free-flow and the cars accelerate as the density drops. Again the slope of these shock waves can be calculated using the same formula $v_{shock} = (q_D - q_U)/(k_D - k_U)$. The difference this time is that traffic flow travels along the fundamental diagram not in a straight line across but many slopes between various points on the curved fundamental diagram. This causes many lines emanating from point x_1 all in a fan shape, called rarefaction. This model implies that the users later on in time will take longer to accelerate as they meet each of the lines. Instead a better approximation is a triangular diagram where the traffic increases abruptly as it would when a driver sees an opening in front of them.

Traffic Congestion

Traffic congestion is a critical problem which happens on roads which make traffic busy because roads full of cars and buses. Traffic congestion challenges traffic flow in urban area and is prevented smooth traffic. A growing urban area creates complex problems in daily life with traffic. Congestion phenomenon's can not be terminated only by applying physical constructing such as: build bridge, motorways and increasing road capacity. It is necessary to build technology system for transportation management which is used for control of the traffic phenomenon. Traffic control systems have direct influence on traffic problems which is help to improve traffic flow and reduce traffic congestion. Usually, traffic jams are caused by many reasons such as incidents, works in roads, roads maintenance. Approximately, traffic congestion occurs at peak times in the morning or evening when people are travelling to work. In addition, transportation systems are one of the main reasons causing traffic congestion in some countries. Traffic roads have a direct effect on traffic jams which relate to small ways use or damaged ways (Katathira et al, 1994)

Firstly, when K. Macmillan invented the first cars had proper shape and wooden frame in 1839 which included many types of advantages such as: motion speed; comfort; control; economy; furniture and safety. The Automobiles industry became more popular and necessary for the life and it aimed to improve car's industrial design and create more cars by companies. In the end of 18th century, the first traffic congestion appeared on the old London Bridge which created a serious problem. After that, the London government decided to find way out for this phenomenon which led to the invention of control system to solve traffic congestion by installing traffic lights on the old London bridge. In addition, the London government added another control system by dividing the London old bridge into two paths, one for vehicles and another for pedestrians (Bellis, 1994).

Transportation problems have been studied by many researchers and many solutions have been proposed to solve it. Improving transportation includes many ways to improve traffic performance in major cities and motorways. The Government is responsible for improving the efficiency of transportation in cities and highways between cities which reduce traffic problems. Transportation blockage has a clear influence on economic situation and it causes increased in live cost.

Traffic congestion is complex phenomenon which is related to a number of cars on the road at the same times which is hinder motion and need extra time to reach destinations. In fact, traffic congestion is usually recurring or non recurring phenomenon. Always, recurring traffic congestion is appeared by daily event in a certain times which is solved reorganize suitable planning. On the other hand, non recurring traffic congestion is unexpected and unusual event cause by an incident which is suddenly reduced road capacity. In addition, road capacity is one of the main reason cause traffic congestion when the number of cars on the road increases than the road capacity.

Causes

Traffic congestion occurs when a volume of traffic or modal split generates demand for space greater than the available street capacity; this point is commonly termed saturation. There are a number of specific circumstances which cause or aggravate congestion; most of them reduce the capacity of a road at a given point or over a certain length, or increase the number of vehicles required for a given volume of people or goods. About half of U.S. traffic congestion is recurring, and is attributed to sheer weight of traffic; most of the rest is attributed to traffic incidents, road work and weather events.

Traffic congestion on Marginal Pinheiros, near downtown São Paulo. According to *Time* magazine, São Paulo has the world's worst traffic jams. Drivers are informed through variable message signs that display the prevailing queue length.

Traffic research still cannot fully predict under which conditions a "traffic jam" (as opposed to heavy, but smoothly flowing traffic) may suddenly occur. It has been found that individual incidents (such as accidents or even a single car braking heavily in a previously smooth flow) may cause ripple effects (a cascading failure) which then spread out and create a sustained traffic jam when, otherwise, normal flow might have continued for some time longer.

Separation of Work and Residential Areas

People often work and live in different parts of the city. Places of work are often located away from housing areas, resulting in the need for people to commute to work. According to a 2011 report published by the United States Census Bureau, a total of 132.3 million people in the country commute between their work and residential areas daily.

Movement from One Part of the City to Another to Obtain or Provide Goods and Services

People may need to move about within the city to obtain goods and services, for instance to purchase goods or attend classes in a different part of the city. Brussels, a city with a strong service economy, has one of the worst traffic congestion in the world, wasting 74 hours in traffic in 2014.

Inadequate Transport Infrastructure and Poor Provision of Transport Services.

This means that the city's transportation facilities are not capable of handling the amount of traffic it receives, such as the lack of alternative routes on roads, and a lack of public transportation where buses and trains are overcrowded and infrequent. In Mumbai, India, trains are often filled to many times their capacity. Buses caught in traffic congestion are often filled with passengers. Therefore, many people turn to driving their own cars to have a more pleasant commute. Thus, many people turn to driving their own cars which can cause a heavier traffic flow.

Mathematical Theories

Some traffic engineers have attempted to apply the rules of fluid dynamics to traffic flow, likening it to the flow of a fluid in a pipe. Congestion simulations and real-time observations have shown that in heavy but free flowing traffic, jams can arise spontaneously, triggered by minor events ("butterfly effects"), such as an abrupt steering maneuver by a single motorist. Traffic scientists liken such a situation to the sudden freezing of supercooled fluid.

Congestion on a street in Taipei consisting primarily of motorcycles

However, unlike a fluid, traffic flow is often affected by signals or other events at junctions that periodically affect the smooth flow of traffic. Alternative mathematical theories exist, such as Boris Kerner's three-phase traffic theory.

Because of the poor correlation of theoretical models to actual observed traffic flows, transportation planners and highway engineers attempt to forecast traffic flow using empirical models. Their working traffic models typically use a combination of macro-, micro- and mesoscopic features, and may add matrix entropy effects, by "platooning" groups of vehicles and by randomising the flow patterns within individual segments of the network. These models are then typically calibrated by measuring actual traffic flows on the links in the network, and the baseline flows are adjusted accordingly.

A team of MIT mathematicians has developed a model that describes the formation of "phantom jams," in which small disturbances (a driver hitting the brake too hard, or getting too close to another car) in heavy traffic can become amplified into a full-blown, self-sustaining traffic jam. Key to the study is the realization that the mathematics of such jams, which the researchers call "jamitons," are strikingly similar to the equations that describe detonation waves produced by explosions, says Aslan Kasimov, lecturer in MIT's Department of Mathematics. That discovery enabled the team to solve traffic-jam equations that were first theorized in the 1950s.

Economic Theories

Congested roads can be seen as an example of the tragedy of the commons. Because roads in most places are free at the point of usage, there is little financial incentive for drivers not to over-use them, up to the point where traffic collapses into a jam, when demand becomes limited by opportunity cost. Privatization of highways and road pricing have both been proposed as measures that may reduce congestion through economic incentives and disincentives. Congestion can also happen due to non-recurring highway incidents, such as a crash or roadworks, which may reduce the road's capacity below normal levels.

India's economic growth has resulted in a massive increase in the number of private vehicles on its roads overwhelming the transport infrastructure. Shown here is a traffic jam in Delhi.

Rapid economic growth in China has resulted in a massive increase in the number of private vehicles in its major cities. Shown here is a traffic jam downtown Haikou City, Hainan Province.

Economist Anthony Downs argues that rush hour traffic congestion is inevitable because of the benefits of having a relatively standard work day. In a capitalist economy, goods can be allocated either by pricing (ability to pay) or by queueing (first-come first-served); congestion is an example of the latter. Instead of the traditional solution of making the "pipe" large enough to accommodate the total demand for peak-hour vehicle travel (a supply-side solution), either by widening roadways or increasing "flow pressure" via automated highway systems, Downs advocates greater use of road pricing to reduce congestion (a demand-side solution, effectively rationing demand), in turn plowing the revenues generated therefrom into public transportation projects.

A 2011 study in *The American Economic Review* indicates that there may be a "fundamental law of road congestion." The researchers, from the University of Toronto and the London School of Economics, analyzed data from the U.S. Highway Performance and Monitoring System for 1983, 1993 and 2003, as well as information on population, employment, geography, transit, and political factors. They determined that the number of vehicle-kilometers traveled (VKT) increases in direct proportion to the available lane-kilometers of roadways. The implication is that building new roads and widening existing ones only results in additional traffic that continues to rise until peak congestion returns to the previous level.

Classification

Qualitative classification of traffic is often done in the form of a six letter A-F level of service (LOS) scale defined in the Highway Capacity Manual, a US document used (or used as a basis for national guidelines) worldwide. These levels are used by transportation engineers as a shorthand and to describe traffic levels to the lay public. While this system generally uses delay as the basis for its measurements, the particular measurements and statistical methods vary depending on the facility being described. For instance, while the percent time spent following a slower-moving vehicle figures into the LOS for a rural two-lane road, the LOS at an urban intersection incorporates such measurements as the number of drivers forced to wait through more than one signal cycle.

Traffic congestion occurs in time and space, i.e., it is a spatiotemporal process. Therefore, another classification schema of traffic congestion is associated with some common spatiotemporal features of traffic congestion found in measured traffic data. Common spatiotemporal empirical features of traffic congestion are those features, which are qualitatively the same for different highways in different countries measured during years of traffic observations. Common features of traffic congestion are independent on weather, road conditions and road infrastructure, vehicular technology, driver characteristics, day time, etc. Examples of common features of traffic congestion are the features [J] and [S] for, respectively, the *wide moving jam* and *synchronized flow* traffic phases found in Kerner's three-phase traffic theory. The common features of traffic congestion can be reconstructed in space and time with the use of the ASDA and FOTO models.

Negative Impacts

A frustrated driver in a traffic jam

Traffic congestion has a number of negative effects:

- Wasting time of motorists and passengers ("opportunity cost"). As a non-productive activity for most people, congestion reduces regional economic health.
- Delays, which may result in late arrival for employment, meetings, and education, resulting in lost business, disciplinary action or other personal losses.
- Inability to forecast travel time accurately, leading to drivers allocating more time to travel "just in case", and less time on productive activities.

- Wasted fuel increasing air pollution and carbon dioxide emissions owing to increased idling, acceleration and braking.
- Wear and tear on vehicles as a result of idling in traffic and frequent acceleration and braking, leading to more frequent repairs and replacements.
- Stressed and frustrated motorists, encouraging road rage and reduced health of motorists
- Emergencies: blocked traffic may interfere with the passage of emergency vehicles traveling to their destinations where they are urgently needed.
- Spillover effect from congested main arteries to secondary roads and side streets as alternative routes are attempted ('rat running'), which may affect neighborhood amenity and real estate prices.
- Higher chance of collisions due to tight spacing and constant stopping-and-going.

Road Rage

Road rage is aggressive or angry behavior by a driver of an automobile or other motor vehicle. Such behavior might include rude gestures, verbal insults, deliberately driving in an unsafe or threatening manner, or making threats. Road rage can lead to altercations, assaults, and collisions which result in injuries and even deaths. It can be thought of as an extreme case of aggressive driving.

The term originated in the United States in 1987–1988 (specifically, from Newscasters at KTLA, a local television station), when a rash of freeway shootings occurred on the 405, 110 and 10 freeways in Los Angeles, California. These shooting sprees even spawned a response from the AAA Motor Club to its members on how to respond to drivers with road rage or aggressive maneuvers and gestures.

Economic Loss

Area	Loss in Billion	Note
USA	$ 305	
UK	£ 40	
NYC	$ 33.7	
LA	$ 19.2	
Manila	$ 18.615	
Bangladesh	$ 11.4	
SF	$ 10.6	
Atlanta	$ 7.1	
Jakarta	5	
Dhaka	US $4.462	
GTHA	$ 3.3	

Positives Impacts

Congestion has the benefit of encouraging motorists to retime their trips so that expensive road space is in full use for more hours per day.

The standard response to congestion is to expand road space somehow, perhaps by widening an existing road or else by adding a new road, bridge or tunnel. However, that could well result in increased traffic flow, otherwise known as induced demand, causing congestion to appear somewhere else. Moreover, Braess' paradox shows that adding road capacity might make congestion worse, even if demand does not increase.

It has been argued that traffic congestion, by reducing road speeds in cities, could reduce the frequency and severity of road accidents.

Countermeasures

Road Infrastructure

Metered ramp on I-894 in Milwaukee, Wisconsin, U.S. The queue of cars waiting at the red light can be seen on the upper portion of the picture.

The HOV lanes in Highway 404 in Southern Ontario are separated by a stripped buffer zone that breaks occasionally to allow vehicles to enter and exit the HOV lane.

- Junction improvements
  - Grade separation, using bridges (or, less often, tunnels) freeing movements from having to stop for other crossing movements
  - Ramp signalling, 'drip-feeding' merging traffic via traffic signals onto a congested motorway-type roadway
  - Reducing junctions
    - Local-express lanes, providing through lanes that bypass junction on-ramp and off-ramp zones
    - Limited-access road, roads that limit the type and amounts of driveways along their lengths
- Reversible lanes, where certain sections of highway operate in the opposite direction on different times of the day(s) of the week, to match asymmetric demand. These pose a potential for collisions, if drivers do not notice the change in direction indicators. This may be controlled by variable-message signs or by movable physical separation
- Separate lanes for specific user groups (usually with the goal of higher people throughput with fewer vehicles)
  - Bus lanes as part of a busway system
  - Express toll lanes
  - HOV lanes, for vehicles with at least three (sometimes at least two) riders, intended to encourage carpooling
    - Slugging, impromptu carpooling at HOV access points, on a hitchhiking or payment basis
    - Market-based carpooling with pre-negotiated financial incentives for the driver

Urban Planning and Design

City planning and urban design practices can have a huge impact on levels of future traffic congestion, though they are of limited relevance for short-term change.

- Grid plans including fused grid road network geometry, rather than tree-like network topology which branches into cul-de-sacs (which reduce local traffic, but increase total distances driven and discourage walking by reducing connectivity). This avoids concentration of traffic on a small number of arterial roads and allows more trips to be made without a car.
- Zoning laws that encourage mixed-use development, which reduces distances between residential, commercial, retail, and recreational destinations (and encourage cycling and walking)
- Carfree cities, car-light cities, and eco-cities designed to eliminate the need to travel by car for most inhabitants.

- Transit-oriented development are residential and commercial areas designed to maximize access to public transport by providing a transit station or stop (train station, metro station, tram stop, or bus stop).

Supply and Demand

Widening works under way on the M25 motorway surrounding London, England to increase the number of lanes

During rush hour, right turns onto the side street shown here are prohibited in order to prevent rat running.

Congestion can be reduced by either increasing road capacity (supply), or by reducing traffic (demand). Capacity can be increased in a number of ways, but needs to take account of latent demand otherwise it may be used more strongly than anticipated. Critics of the approach of adding capacity have compared it to "fighting obesity by letting out your belt" (inducing demand that did not exist before). For example, when new lanes are created, households with a second car that used to be parked most of the time may begin to use this second car for commuting. Reducing road capacity has in turn been attacked as removing free choice as well as increasing travel costs and times, placing an especially high burden on the low income residents who must commute to work.

Increased supply can include:

- Adding more capacity at bottlenecks (such as by adding more lanes at the expense of hard shoulders or safety zones, or by removing local obstacles like bridge supports and widening tunnels)

- Adding more capacity over the whole of a route (generally by adding more lanes)
- Creating new routes
- Traffic management improvements

Reduction of demand can include:

- Parking restrictions, making motor vehicle use less attractive by increasing the monetary and non-monetary costs of parking, introducing greater competition for limited city or road space. Most transport planning experts agree that free parking distorts the market in favour of car travel, exacerbating congestion.
- Park and ride facilities allowing parking at a distance and allowing continuation by public transport or ride sharing. Park-and-ride car parks are commonly found at metro stations, freeway entrances in suburban areas, and at the edge of smaller cities.
- Reduction of road capacity to force traffic onto other travel modes. Methods include traffic calming and the shared space concept.
- Road pricing, charging money for access onto a road/specific area at certain times, congestion levels or for certain road users
  - "Cap and trade", in which only licensed cars are allowed on the roads. A limited quota of car licences are issued each year and traded in a free market fashion. This guarantees that the number of cars does not exceed road capacity while avoiding the negative effects of shortages normally associated with quotas. However, since demand for cars tends to be inelastic, the result are exorbitant purchase prices for the licenses, pricing out the lower levels of society, as seen Singapore's Certificate of Entitlement scheme.
  - Congestion pricing, where a certain area, such as the inner part of a congested city, is surrounded with a cordon into which entry with a car requires payment. The cordon may be a physical boundary (i.e., surrounded by toll stations) or it may be virtual, with enforcement being via spot checks or cameras on the entry routes. Major examples are Singapore's electronic road pricing, the London congestion charge system, Stockholm congestion tax and the use of High-occupancy toll lanes, predominately in North America.
- Road space rationing, where regulatory restrictions prevent certain types of vehicles from driving under certain circumstances or in certain areas.
  - Number plate restrictions based on days of the week, as practiced in several large cities in the world, such as Athens, Mexico City, Manila and São Paulo. In effect, such cities are banning a different part of the automobile fleet from roads each day of the week. Mainly introduced to combat smog, these measures also reduce congestion. A weakness of this method is that richer drivers can purchase a second or third car to circumvent the ban.
  - Permits, where only certain types of vehicles (such as residents) are permitted to enter a certain area, and other types (such as through-traffic) are banned. For example, Bertrand Delanoë, the mayor of Paris, has proposed to impose a complete ban on motor vehicles in the city's inner districts, with exemptions only for residents, businesses, and the disabled.

Bike lane constructed in congested areas to encourage use of the alternative transportation

- Policy approaches, which usually attempt to provide either strategic alternatives or which encourage greater usage of existing alternatives through promotion, subsidies or restrictions.
    - Incentives to use public transport, increasing modal shares. This can be achieved through infrastructure investment, subsidies, transport integration, pricing strategies that decrease the marginal cost/fixed cost ratios, improved timetabling and greater priority for buses to reduce journey time e.g. bus lanes or bus rapid transit .
    - Cycling promotion through legislation, cycle facilities, subsidies, and awareness campaigns. The Netherlands has been pursuing cycle friendly policies for decades, and around a quarter of their commuting is done by bicycle.
    - Promotion of more flexible work place practices. For example, a flexible workplaces pilot was undertaken in Brisbane, Australia during 2009 to test the applicability of a voluntary travel behaviour change program to achieve transport system outcomes, particularly as they related to managing congestion, either through mode shift or peak spreading. During the one-month Pilot, amongst almost 900 Brisbane CBD workers across 20 private and public sector organisations, shifts of more than 30% out of the morning and afternoon peak travel was recorded.
    - Telecommuting encouraged through legislation and subsidies.
    - Online shopping promotion, potentially with automated delivery booths helping to solve the last mile problem and reduce shopping trips made by car.

Traffic Management

Use of so-called Intelligent transportation system, which guide traffic:

- Traffic reporting, via radio, GPS and mobile apps, to advise road users
- Variable message signs installed along the roadway, to advise road users
- Navigation systems, possibly linked up to automatic traffic reporting
- Traffic counters permanently installed, to provide real-time traffic counts

- Convergence indexing road traffic monitoring, to provide information on the use of highway on-ramps
- Automated highway systems, a future idea which could reduce the safe interval between cars (required for braking in emergencies) and increase highway capacity by as much as 100% while increasing travel speeds
- Parking guidance and information systems providing dynamic advice to motorists about free parking
- Active Traffic Management system opens up UK motorway hard shoulder as an extra traffic lane; it uses CCTV and VMS to control and monitor the traffic's use of the extra lane.

Traffic congestion detector in Germany

Other Associated

- School opening times arranged to avoid rush hour traffic (in some countries, private car school pickup and drop-off traffic are substantial percentages of peak hour traffic).
- Considerate driving behaviour promotion and enforcement. Driving practices such as tailgating and frequent lane changes can reduce a road's capacity and exacerbate jams. In some countries signs are placed on highways to raise awareness, while others have introduced legislation against inconsiderate driving.
- Visual barriers to prevent drivers from slowing down out of curiosity (often called "rubbernecking" in the United States). This often includes accidents, with traffic slowing down even on roadsides physically separated from the crash location. This also tends to occur at construction sites, which is why some countries have introduced rules that motorway construction has to occur behind visual barrier
- Speed limit reductions, as practiced on the M25 motorway in London. With lower speeds allowing cars to drive closer together, this increases the capacity of a road. Note that this measure is only effective if the interval between cars is reduced, not the distance itself. Low intervals are generally only safe at low speeds.

- Lane splitting/filtering, in which some jurisdictions allow motorcycles, scooters and bicycles to travel in the space between cars, buses, and trucks.
- Reduction of road freight avoiding problems such as double parking with innovative solutions including cargo bicycles and Gothenburg's Stadsleveransens.

Traffic Reporting

Survey of how people consume traffic information reveals that technology is rapidly changing the role of TV stations versus digital platforms. The research should also serve as a wakeup call for broadcasters to get onboard with digital technology quickly, or risk seeing their consumers going elsewhere.

First, it's important to understand that technology has changed how traffic data is collected. The traditional approach had been to have a helicopter or two in the air reporting live and feeding information back to the station. You had drivers out on the road, especially during weather events. And you maintained a traffic operations center to listen to police scanners, take calls from people, and watch available camera feeds. It was a model that worked in a pre-digital world but in recent years has become obsolete and unsustainable as business models have shifted.

With mobile technology, a whole new data source evolved, one rich with real-time traffic information. GPS enabled probes tell us how fast traffic is moving, how long it will take to drive a section of road, average speeds, whether traffic is getting better or worse, and so on. It's also now easier than ever to get access to incident data from a wide variety of sources – web and mobile sites, social media - even cameras and police scanners can be accessed from any mobile device.

Reaching the Changing Consumer

Mobile technology has also changed how people consume information. The research shows that smartphones and the Internet are seriously competing with TV as the consumers' choice for traffic information, especially among younger people. In fact on weekday afternoons and evenings, smartphones are preferred to TV; of course this makes sense given that people are probably leaving work for the drive home. In the early weekday morning, TV is still preferred but smartphones are gaining ground.

Maybe most interesting of all is this finding from the research: consumers prefer to use smartphone apps from their local TV station and they like to visit the local station's web site too. This suggests there's an important opportunity out there, for those TV stations that can stay ahead of the curve.

Methods of Gathering Many Information

There are several methods in use today to gather traffic speed and incident info, ranging from professional reporters, to GPS crowdcrowding to combinations of both methods.

- INRIX uses its network of over 175 million vehicles and devices to gather speed data from mobile phones, trucks, delivery vans, and other fleet vehicles equipped with GPS locator devices including smart phones and Ford SYNC and Toyota Entune and much of Europe, South America, and Africa.

- Google Traffic works by crowdsourcing the GPS information from phone users. By calculating the speed of users along a stretch of road, Google is able to generate a live traffic map. Its subsidiary, Waze, also allows users to report directly via a smartphone app.
- TomTom Traffic uses crowd-sourced data from mobile phone users, along with data from traditional sources such as induction loops and traffic cameras.
- Monitoring police radio frequencies. Some radio stations have agreements with states' highway patrol that permit a direct connection with a law enforcement computer. This enables real-time information gathering of the latest accident reports to highway patrol divisions.
- Many areas have helicopters to overfly accident scenes and other areas of high traffic volume. For example, by the company Global Traffic Network.
- traffic cameras
- Giditraffic is an online social service which employs crowd sourcing as its primary means of providing real-time traffic updates to subscribers. The service is delivered free of charge.
- RoadPal uses crowd-sourced data from mobile users as well as the social media to provide users with traffic information of places of interest to them.
- Roadside speed sensors, either infra-red sensors for spot measurements or automatic number plate recognition for measuring speed between two sites (Trafficmaster have a network of sensors on motorways and trunk roads in the UK).

Methods of Transmitting Information

- GPS units
- Smartphones
- Radio via voice RDS, and TA
- Electronic road signs
- 5-1-1 traffic information phone line or similar.
- Television and web

Providers

- INRIX develops and distributes INRIX Traffic, a free mobile application, and also provides reporting services to a variety of local television stations.
- NAVTEQ provides data used in a wide range of applications, including automotive navigation systems for many car makers. Most clients use Navteq to provide traffic reports in major metropolitan areas throughout North America. NAVTEQ partners with third-party

agencies and companies to provide its services for portable GPS devices made by Garmin, Lowrance, NDrive and web-based applications such as Yahoo! Maps, Bing Maps, and Nokia Maps. XM Satellite Radio and Sirius Satellite Radio use NAVTEQ data to show traffic information on navigation systems.

NAVTEQ's media services was spun out to form Radiate Media in 2011, which subsequently merged with Global Traffic Network in 2016, forming US Traffic Network (USTN).

- Tele Atlas, a subsidiary of TomTom. delivers digital maps and other dynamic content for navigation and location-based services, including personal and in-car navigation systems, and provides data used in a wide range of mobile and Internet map applications.
- Google Maps uses a variety of governmental and private traffic reporting organizations to provide information, along with its Waze subsidiary, which uses crowdsourcing to provide observed traffic conditions.

Road Traffic Control

Traffic control is supervision of the movement of people, goods, or vehicles to ensure efficiency and safety.

One of the principal challenges in traffic control is to accommodate the traffic in a safe and efficient way. Efficiency can be thought of as a measure of movement levels relative to the objectives for a particular transportation system and the finances required for its operation. For example, a railroad can be thought of as efficient if it can accommodate the travel requirements of its customers at the least cost. It will be thought of as inefficient if an alternative (*e.g.*, a trucking service) can also meet customer needs but at a lower cost.

Safety, the management of traffic to reduce or eliminate accidents, is the other critical reason for traffic control. An airline pilot needs to be warned of high winds at the destination airport just as an automobile driver needs to be warned of a dangerous curve or intersection ahead. Traffic control has as its principal objective to manage the movement of people and goods as efficiently and safely as possible. The dual objectives, however, frequently conflict or, at least, compete. For example, there are frequent cases in which commercial airlines are held on the ground at their originating airport until they receive a clearance to land at a destination. The clearance is given only when the destination airport determines that the number of airplanes expected to arrive at a particular time is small enough that local air traffic controllers can assist the plane in landing without overtaxing their human limitations and compromising safety.

In road traffic, intersections with traffic lights (*i.e.*, green, amber, and red indications) will often add a separate lane with a lighted green arrow to allow left turns with no opposing traffic. This frequently results in longer nongreen periods at the intersection, causing an increased delay and a reduction in efficiency and mobility. Traffic control will always be burdened with seeking to satisfy the frequently conflicting goals of safety and mobility.

Safety is not the exclusive concern of the traffic control community. Nearly every transportation mode has organizations that regulate operators through a series of licensing procedures, sanctions for inappropriate operating practices, and requirements for continuing training to retain certification to operate. Examples include federal aviation authorities that oversee pilot training (*e.g.*, the U.S. Federal Aviation Administration); road agencies that administer driver's licenses may exist at the provincial level (as in Canada) or at the national level (as is more common in Europe). Transportation safety management is thus accomplished through a complex set of interactions between different agencies at different levels (*e.g.*, national, regional or state, and local) using both formal legal requirements and administrative actions. The following discussion will necessarily focus on safety concerns that evolve from and are a component of the traffic control function.

Overview

Road Traffic control is an outdoors occupation, night or day for long hours in all weathers, and is considered a dangerous occupation due to the high risk of being struck by passing vehicles. Safety equipment is vitally important. Fatigue is a big issue, as tired TC's may forget to watch their traffic, or may inadvertently turn their "Stop bats" to the "Slow" position. Many drivers are annoyed by the disruption to their route, and some are sufficiently antisocial as to aim at traffic controllers. Other drivers simply don't pay enough attention to the road, often from using their mobile (cell-) phones, or because they are tired from a night shift at work. Not a few are exceeding the posted speed limit.

Traffic cones on the A45 in South Coventry (U.K.).

Typically, a worksite will be set up with warning signage well in advance of the actual work area. This may involve (in Australia) "*Roadworks Ahead*", temporary speed restrictions, "*Worker Symbolic*" (a stylized workman with a pile of rubble, black silhouette on a retroflective orange background), "*Reduce Speed*", "*Lane Status*" boards (indicating that some lanes on a multilane roadway will be closed), "*Prepare to Stop*" and advisory signs telling what is happening (e.g. *Water Over Road, Trucks Entering*, and *Power Line Works Ahead*). If lanes have been closed, large flashing arrows (Arrow Boards) on trailers may be utilized to give motorists hundreds of meters warning to move over. Motorists will be advised they are leaving a worksite by speed reinstatement or "*End Roadworks*" signs.

The worksite will usually involve closing a part of the road for the work area. How this is done depends on the type of road: on a multi-lane road, one or more lanes will be closed off and traffic merged into the remaining lane(s), using cones and "*Chevron*" signs and Arrow Boards to guide motorists. On a wide road (more than 3 meters per lane in Australia), traffic could be "diverted" around the work area by using cones to define a new road centerline and another line of cones around the work area. Sometimes, it is necessary to close a road and detour traffic.

Traffic diverted around work area in Kwinana, Western Australia

Often, the road is not wide enough to permit opposing streams of traffic past the work area. Then it is necessary to use "*Stop/Slow Paddles*", where each stream is allowed past the work area in turn. On an intersection, this may involve four or more streams. At signalised intersections, it may be necessary to have the traffic lights disabled.

Sometimes on dual carriageways, it is necessary to divert one carriageway onto the opposing carriageway, forming a "contraflow". This cannot be done "on the fly", as high-speed (100+km/h), high-volume (500 - 1000+ vehicles per hour) traffic is involved, generating a huge risk to workers. In this case advisory signs will be erected weeks or even months in advance, and new lanes defined by bollards anchored firmly to the road-base will be installed, usually at night when traffic is expected to be minimal. Programmable *Variable Message Signs* may be utilized at strategic locations to inform motorists.

International perspectives

Australia

Traffic cones placed around a rollover accident in Sydney.

Traffic control is governed by the Australian Standard AS 1742.3 – 2009, and by State variations. Risk management is regulated under AS/NZS 4360:1999. Traffic controllers are required to wear high-visibility clothing which meets the Australian Standard AS/NZS 4602:1999.

Personal safety is emphasised in all Australian training. This ranges from proper clothing to learning appropriate behaviour (for example, always face oncoming traffic). Clothing is considered part of *PPE*—Personal Protective Equipment—which includes steel-capped boots, sunscreen, broad-brim hats, gloves and sunglasses.

The traffic control process usually starts with a traffic control plan(traffic control plan). A traffic control crew may consist of one person running a simple diversion or closure of a cul-de-sac, up to multiple two- or three-person crews for a complex task. One example of such a complex task is the transport of very wide loads taking all available roadspace, over several kilometers, usually on an arterial road or highway. In these cases, the affected roads can be closed or contraflowed for the entire day, creating enormous disruption to motorists. Management of the event involves monitoring and closing all intersections, *Stop/Slow* to work traffic streams through partially closed intersections, and detours. The amount of signage required can be staggering, needing some hours to put in place. Normally a single two-person crew with one ute is sufficient for most jobs.

Traffic control ute at worksite

Not all TC's are employed by dedicated traffic management companies. Many construction companies and government authorities employ their own traffic management. In these cases, TC's will work in other capacities when traffic management is not required.

Traffic control is generally not seen as a career for young people, but rather as a stop-gap while something better is sought. However, older people are often valued by employers for their life-experience, and find that the relatively light manual labour compensates for the discomforts and rigours of the job. There is a career path, but it is dictated by one's own ability and willingness to work.

Western Australia

Accreditation course standards and variations to the Australian Standards are regulated by Main Roads Western Australia (MRWA), part of the Ministry of Planning and Infrastructure.

In Western Australia, use of the *Stop/Slow* bat is authorised under Regulation 83 of the Road Traffic Code 2000—it is an offence to disobey a traffic controller's bat, punishable by 3 demerit points and 3 penalty units (about A$175). Other States have similar provisions.

Cone taper for a "slow lane" closure showing small chevron (shifter), 40 km/h repeater, chevron and arrow-board

Traffic controllers must be accredited in *Basic Worksite Traffic Management BCC3028A* and the *Worksite Traffic Controller Course BCC1014A*. These qualifications must be renewed after three years, and a refresher course is necessary. The courses take about 4 hours each, and are designed as inductions to on-the-job training.

The *Advanced Worksite Traffic Management (AWTM)* requires two years experience as a qualified TC as a minimum prerequisite, and must also be renewed after three years. *Roadworks Traffic Managers* can be accredited with a minimum of five years experience, current "Road Safety Auditor" accreditation and current AWTM accreditation. This qualification is also valid for three years.

All employers require drug screening at least annually and often randomly, and many others require daily blood/alcohol tests; some require police clearance checks. Zero-tolerance is universal. Traffic controllers are usually employed on a casual basis, with wages around A$16 to A$25 per hour.

Canada

British Columbia

In BC, WorkSafeBC regulates the training of Traffic Control Persons (TCPs), stating that TCPs must be trained in a manner acceptable to the Board. This ensures a high level of training for this high-risk occupation. Currently, the only acceptable course in the province is a two-day session which includes theory and practical components. Royal Canadian Mounted Police (RCMP) Superintendent Derek Cooke of Langley believes that the RCMP should not perform the function of road traffic control to cater to events in support of for-profit corporations unless the municipal government has coordinated or is in support of the event.

Nova Scotia

In Nova Scotia training is regulated by the Nova Scotia Department of Transportation and Infrastructure Renewal. There is a one-day course for TCPs and a two-day course for Temporary Workplace Signers. Signers are responsible for the setup of signs, cones etc., and making sure the setup complies with the NS Temporary Workplace Traffic Control Manual

Newfoundland & Labrador (NL)

All flag persons – or traffic control persons (TCP) – in Newfoundland and Labrador are now required to complete a TCP training course approved by the Workplace Health, Safety and Compensation Commission.

Proper traffic control is critical for the safety of workers, drivers and the general public. Without training by a Commission approved training provider, workers are not permitted to work as a TCP on our province's roads.

The Commission's Traffic Control Person (TCP) Certification Training Standard establishes criteria for TCP training providers and trainers.

TCP training providers must apply and be approved by the Commission to deliver TCP certification training.

The delivery of training prepares the TCP to perform traffic control in a safe and competent manner by providing them with the knowledge and skills to work safely, consistent with industry and legislative standards.

Traffic Control Person (TCP) Certification Training has an expiry date of 3 years, upon which the course must be completed again for renewal.

United Kingdom

A small traffic control scheme on a residential UK street

Traffic management in the UK is overseen by the Department for Transport but each country within the union has their own transport regulator.

All countries within the UK adhere to the same standards for most traffic control and temporary modifications to traffic and pedestrian control, such as street works and road works (sometimes referred to as the "Red Book"), entitled Safety at Street Works and Road Works - A Code of Practice

Before permission for any temporary traffic control is given, a Traffic Management Plan must be submitted to the local planning office of the affected district.

England and Wales:

Responsibility is with the Department for Transport. The code of practice for Street and Road works is issued by the Secretary of State for Transport and Welsh Ministers under Section 65 of the New Roads and Street Works Act 1991 (NRSWA) and section 174 of the Highways Act 1980.

Scotland:

Traffic management in Scotland is handled by Traffic Scotland and Transport Scotland. The code of practice for Street and Road works is issued by the Scottish ministers under section 124 of the NRSWA.

Northern Ireland:

Responsibility is with the Department for Regional Development. The code of practice for Street and Road works is issued by the same Department for Regional Development of the Street Works (Northern Ireland) Order 1995 and Road Traffic Regulation Order (Northern Ireland) 1997.

Some of the most iconic images of a traffic controller in the UK, is the Crossing Guard (normally referred to as the "lollipop man" or "lollipop woman") who aids children in road crossing on their journey to school or the Stop / Go Marshall, (Traffic Guard) at temporary road works.

United States

Although the Federal Highway Administration specifies standards and guidelines through the MUTCD which apply to the usage of traffic control equipment, individual state and local agencies often provide additions or slight variations to these standards.

Traffic controller on Michigan Avenue, Chicago

The transportation system in the United States is complex and extensive. Traffic volumes, types of vehicles, driving styles, population density, speed limits, and many other factors vary dramatically from one region to the next. As a result, highway traffic control measures (including type of equipment and implementation), are not strictly consistent. Federal Guidelines do not address certification methods for traffic controllers, flaggers, or other personnel responsible for traffic control. This responsibility is managed on a state or local agency level, and therefore certification requirements are not consistent and are administered locally. Safety standards (irrespective of traffic control) are mandated by OSHA as well as state-level occupational safety departments.

A construction traffic control company operates in the same basic way as any other construction company. Companies submit a bid for a job, the lowest bid is accepted (except in the case of disadvantaged companies), and the labor is provided to the contractor or agency in charge. Typically speaking, flaggers work in groups of 5 to 10 under a TCS, or Traffic Control Supervisor. The TCS is responsible for placing the flaggers correctly, ensuring that they receive the proper breaks and supervision, and placing the advance warning signs (such as Road Work Ahead, One Lane Road Ahead, and Uneven Lanes). Flaggers are the second line of attention (after the warning signs) for drivers. They are the first people in the work zone to deal with opposing traffic.

While construction traffic control in the U.S. used to be a widely unionized profession, it is now dominated by private business and wages are not controlled by the union.

While MUTCD and other standard practices have been instilled in traffic control workers for many years, workers still need to keep up to date on all the new regulations.

Some road traffic control systems have started to optimize multiple traffic modes, including vehicles and pedestrians.

Traffic Optimization

Traffic modeling has been of interest to mathematicians since the 1950s. Research in the area has only grown as road traffic control presents an ever-increasing problem.

Generally, models for traffic flow in road networks are time-dependent and continuous, that is, they describe traffic by a continuum rather than as individual drivers or cars. These macroscopic models describe the temporal and spatial evolution of traffic density without predicting traffic patterns of individuals. In addition to macroscopic models based on continuous densities, microscopic approaches like particle models or cellular automata are also used to model traffic.

Traffic at Song-shou and Song-chih Intersection in Taipei, Taiwan

Most existing continuous models consider unidirectional traffic; thus, the traffic density depends only on a single spatial dimension. The governing equations in this class of macroscopic models are inspired by gas dynamics equations.

A lot of recent work has focused on traffic intersections, which constitute a building block of larger road networks. Here, models generally aim to either minimize travel time of individual drivers, or maximize the total traffic flow at a given intersection.

"Mathematical equations that model vehicular traffic, similar to fluid flow, are able to capture nonlinear phenomena, such as, traffic jam formation," explains author Simone Göttlich. "Traffic lights are a necessary tool to redirect the traffic flow within road networks and therefore offer the potential to mitigate congestion even for high traffic volumes based on mathematical insights.

Usually, traffic optimization models, which are based on cell-transmission, fluid, mixed-integer formulations, and heuristics, attempt to find an optimal cycle length of green and red phases for traffic lights.

Need for Traffic Optimization

Texas Transportation Institute estimates travel delays of between 17–55 hours of delay per person per year relating to congestion on the streets. Traffic device optimization hence becomes a significant aspect of operations.

Techniques

Several techniques exist to reduce delay of traffic. Generally the algorithms attempt to reduce delays (user time), stops, exhaust gas emissions, or some other measure of effectiveness. Many optimization software are geared towards pre-timed coordinated systems.

Normally optimization of signals along a road is a challenging and expensive task, because the sources for traffic monitoring have been limited to inductive loops, cameras or manually counting. However, due to recent advances in information technology, portable devices with Bluetooth and Wi-Fi communication are becoming more common,enabling real-time continuous traffic monitoring and adjustments to traffic signal timing.

By placing sensors along roads, tracking Bluetooth and Wi-Fi devices in passing vehicles, the solution is able to accurately detect and record how long it takes a car to drive along a corridor, segment by segment and in total. This provides historic data for traditional timing methods but also enables real-time feedback to changes in signal programs along with the ability to continuously detect traffic levels and travel time to trigger transitions among programs.

Real Time Traffic Control

Several systems are capable of monitoring the traffic arrivals and adjusting timings based on the detected inputs. Traffic Detectors may range from Metal Detectors to Detectors that use Image Detection. Metal detectors are the most popular in use. Image detection devices exhibit numerous problems including degradation during bad weather and lighting.

Traffic actuated signal systems use detectors to adjust timing for:

1. Only the main street - semi-actuated system
2. Both main and cross streets - fully actuated system.

The above method is primitive real-time signal optimization at best. This method will optimize one traffic signal at a time. However, in the real world, a motorist's commute involves driving through multiple signals. Thus, multiple traffic signals need to be collectively synchronized in order to be effective. One such system that has gained significant popularity in the United States is InSync.

Criticism

It has been suggested that the benefits of traffic optimization have never been scientifically justified. It inherently favors motorized traffic over alternate modes such as pedestrians, bicyclists, and transit users and may promote more auto use. It is suggested that an alternate approach could involve traffic calming, and a conceptual focus on the movement of people and goods rather than vehicles.

It can be argued that Traffic optimization inherently calms traffic due to discouraging speeding and limits acceleration and deceleration thus reducing the noise pollution produced by vehicles.

Traffic Light Control and Coordination

Traffic jams and suboptimal traffic flows are ubiquitous in our modern societies, and they create enormous economic losses each year. Delays at traffic lights alone contribute roughly 10 percent of all delays in US traffic. As most traffic light scheduling systems currently in use are static, set up by human experts rather than being adaptive, the interest in machine learning approaches to this problem has increased in recent years. Reinforcement learning approaches are often used in these studies, as they require little pre-existing knowledge about traffic flows. Some distributed constraint optimization approaches have also been used, but focus on cases where the traffic flows are known. This paper presents a preliminary comparison between these two classes of optimization methods in a complex simulator, with the goal of eventually producing real-time algorithms that could be deployed in real-world situations.

Traffic in a city is very much affected by traffic light controllers. When waiting for a traffic light, the driver looses time and the car uses fuel. Hence, reducing waiting times before traffic lights can save our European society billions of Euros annually. To make traffic light controllers more intelligent, we exploit the emergence of novel technologies such as communication networks and sensor networks, as well as the use of more sophisticated algorithms for setting traffic lights. Intelligent traffic light control does not only mean that traffic lights are set in order to minimize waiting times of road users, but also that road users receive information about how to drive through a city in order to minimize their waiting times. This means that we are coping with a complex multi-agent system, where communication and coordination play essential roles. Our research has led to a novel system in which traffic light controllers and the behaviour of car drivers are optimized using machine-learning methods.

Phases and Stages

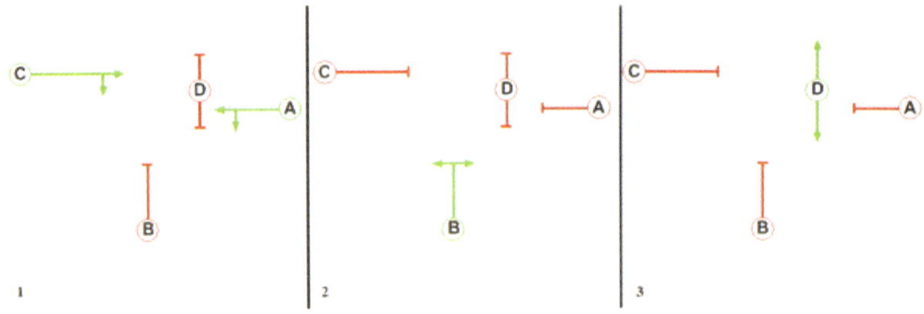

This junction has three vehicle phases (A, B and C) and a pedestrian phase (D). The phases operate together in three stages (1, 2 and 3). Moving phases are shown in green and stopped phases are shown in red.

Traffic controllers use the concept of phases, which are directions of movement grouped together. For instance, a simple T-junction may have three vehicle movement phases, one for each arm of the junction. There may be additional phases for other movements such as pedestrians, cyclists, bus lanes or tramways.

A stage is a group of non-conflicting phases which move at the same time.

Traffic Controller Systems

A traffic signal is typically controlled by a controller mounted inside a cabinet. Some electro-mechanical controllers are still in use (New York City still had 4,800 as of 1998, though the number is lower now due to the prevalence of the signal controller boxes). However, modern traffic controllers are solid state. The cabinet typically contains a power panel, to distribute electrical power in the cabinet; a detector interface panel, to connect to loop detectors and other detectors; detector amplifiers; the controller itself; a conflict monitor unit; flash transfer relays; a police panel, to allow the police to disable the signal; and other components.

Computerized traffic control box

In the United States, controllers are standardized by the NEMA, which sets standards for connectors, operating limits, and intervals. The TS-1 standard was introduced in 1976 for the first generation of solid-state controllers.

Solid state controllers are required to have an independent *conflict monitor unit*(CMU), which ensures fail-safe operation. The CMU monitors the outputs of the controller, and if a fault is detected, the CMU uses the flash transfer relays to put the intersection to *FLASH*, with all red lights flashing, rather than displaying a potentially hazardous combination of signals. The CMU is programmed with the allowable combinations of lights, and will detect if the controller gives conflicting directions a green signal, for instance.

In the late 1990s, a national standardization effort known as the Advanced transportation controller (ATC) was undertaken in the United States by the Institute of Transportation Engineers. The project attempts to create a single national standard for traffic light controllers. The standardization effort is part of the National Intelligent transportation system program funded by various highway bills, starting with ISTEA in 1991, followed by TEA-21, and subsequent bills. The controllers will communicate using *National Transportation Communications for ITS Protocol* (NTCIP), based on Internet Protocol, ISO/OSI, and ASN.1.

Battery backups installed in a separate cabinet from the traffic controller cabinet on the top.

Traffic lights must be instructed when to change stage and they are usually coordinated so that the stage changes occur in some relationship to other nearby signals or to the press of a pedestrian button or to the action of a timer or a number of other inputs.

Battery backup

In the areas that are prone to power interruptions, adding battery backups to the traffic controller systems can enhance the safety of the motorists and pedestrians. In the past, a larger capacity of uninterruptible power supply would be required to continue the full operations of the traffic signals using incandescent lights. The cost for such system would be prohibitive. After the newer generations of traffic signals that use LED lights which consume 85-90% less energy, it is now possible to incorporate battery backups into the traffic light systems. The battery backups would be installed in the traffic controller cabinet or in their own cabinet adjacent to the controller.

The battery backups can operate the controller in emergency mode with the red light flashing or in fully functional mode. In 2004, California Energy Commissionrecommended to have local governments to convert their traffic lights to LEDs with battery backups. This would lower the energy consumption and enhance the safety at major intersections. The recommendation was for a system which provides fully functional traffic signals for two hours after the power outage. Then the signals will have flashing red lights for another two hours.

Fixed Time Control

Pedestrian traffic signal in Taiwan, featuring a "Walking green man" below a countdown display where the "Red Man" once stood.

In traffic control, simple and old forms of signal controllers are what are known as electro-mechanical signal controllers. Unlike computerized signal controllers, electro-mechanical signal controllers are mainly composed of movable parts (cams, dials, and shafts) that control signals that are wired to them correctly. Aside from movable parts, electrical relays are also used. In general, electro-mechanical signal controllers use dial timers that have fixed, signalized intersection time plans. Cycle lengths of signalized intersections are determined by small gears that are located within dial timers. Cycle gears, as they are commonly known, range from 35 seconds to 120 seconds. If a cycle gear in a dial timer results in a failure, it can be replaced with another cycle gear that would be appropriate to use. Since a dial timer has only one signalized intersection time plan, it can control phases at a signalized intersection in only one way. Many old signalized intersections still use electro-mechanical signal controllers, and signals that are controlled by them are effective in one way grids where it is often possible to coordinate the signals to the posted speed limit. They are however disadvantageous when the signal timing of an intersection would benefit from being adapted to the dominant flows changing over the time of the day.

Coordinated Control

Attempts are often made to place traffic signals on a coordinated system so that drivers encounter a green wave, a long string of green lights (the technical term is progression). The distinction between coordinated signals and synchronized signals is very important. Synchronized signals all change at the same time and are only used in special instances or in older systems. Coordinated (progressed) systems are controlled from a master controller and are set up so lights "cascade" (progress) in sequence so platoons of vehicles can proceed through a continuous series of green lights. A graphical representation of phase state on a two-axis plane of distance versus time clearly shows a "green band" that has been established based on signalized intersection spacing and expected vehicle speeds. In some countries (e.g. Germany, France and the Netherlands), this "green band" system is used to limit speeds in certain areas. Lights are timed in such a way that motorists can drive through without stopping if their speed is lower than a given limit, mostly 50 km/h (30 mph) in urban areas. This system is known as "grüne Welle" in German, "vague verte" in French, or "groene golf" in Dutch (English: "green wave"). Such systems were commonly used in urban areas of the United States from the 1940s, but are less common today. In the UK, Slough in Berkshire had part of the A4 experimented on with this. Many US cities set the green wave on two-way streets to operate in the direction more heavily traveled, rather than trying to progress traffic in both directions. But the recent introduction of the flashing yellow arrow makes the lead-lag signal, an aid to progression, available with protected/permissive turns.

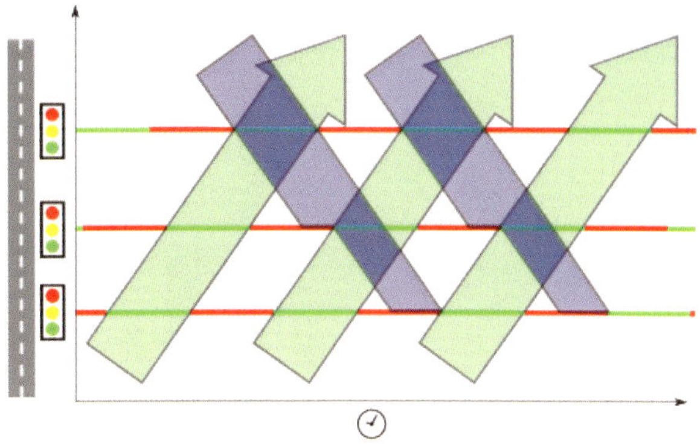

Diagram demonstrating that when traffic lights are synchronised for traffic travelling in one direction (green arrows), the traffic in the other direction is not necessarily synchronised (blue arrows).

In modern coordinated signal systems, it is possible for drivers to travel long distances without encountering a red light. This coordination is done easily only on one-way streets with fairly constant levels of traffic. Two-way streets are often arranged to correspond with rush hours to speed the heavier volume direction. Congestion can often throw off any coordination, however. On the other hand, some traffic signals are coordinated to prevent drivers from encountering a long string of green lights. This practice discourages high volumes of traffic by inducing delay yet preventing congestion. Speed is self-regulated in coordinated signal systems; drivers traveling too fast will arrive on a red indication and end up stopping, drivers traveling too slowly will not arrive at the next signal in time to utilize the green indication. In synchronized systems, however, drivers will often use excessive speed in order to get through as many lights as possible.

This traffic light in Khobar, Saudi Arabia is video camera-actuated (just above the vertically-aligned lenses) and also shows the seconds remaining to change to the next state (in the leftmost horizontally-aligned lens)

More recently even more sophisticated methods have been employed. Traffic lights are sometimes centrally controlled by monitors or by computers to allow them to be coordinated in real time to deal with changing traffic patterns. Video cameras, or sensors buried in the pavement can be used to monitor traffic patterns across a city. Non-coordinated sensors occasionally impede traffic by detecting a lull and turning red just as cars arrive from the previous light. The most high-end systems use dozens of sensors and cost hundreds of thousands of dollars per intersection, but can very finely control traffic levels. This relieves the need for other measures (like new roads) which are even more expensive.

Benefits include:

- Increasing the traffic handling capacity of roads
- Reducing collisions and waiting time for both vehicles and pedestrians
- Encouraging travel within the speed limit to meet green lights
- Reducing unnecessary stopping and starting of traffic - this in turn reduces fuel consumption, air and noise pollution, and vehicle wear and tear
- Reducing travel time
- Reducing driver frustration and road rage

Examples:

- New York City: 7,660 (of a total of 12,460) signalized intersections are controlled by a central computer network and monitored by traffic management centers.
- Toronto: 83% of its signals are controlled by the Main Traffic Signal System (MTSS). 15% also use the SCOOT (Split Cycle and Offset Optimization Technique), an adaptive signal control system.
- Sydney: 3,400 traffic signals co-ordinated by the Sydney Co-ordinated Adaptive Traffic System (SCATS). Designed and developed by RTA, the system was first introduced in 1963 and progressively developed since then. By October 2010, SCATS was licensed to 33,200 intersections in 144 cities across 24 countries worldwide, including Singapore, Hong Kong, Dublin, Tehran and Minneapolis and Detroit.
- Melbourne: 3,200 traffic lights across Victoria, including regional areas such as Geelong and Ballarat, using SCATS. Some 500 intersections also have tram and bus priority.
- Adelaide: 580 sets of coordinated traffic lights throughout the metropolitan region managed by the Adelaide Coordinated Traffic Signal (ACTS) System.

Adaptive Control

- Meadowlands Adaptive Signal System for Traffic Reduction (MASSTR) - New Jersey Meadowlands Commission monitors in their Lyndhurst administration building this “intelligent transportation system”, the first of its kind in the state, with traffic controlled intersections and vehicle detectors in the Meadowlands. As of 2013, it is in operation and plans to cover 128 intersections by 2014. To reduce delays due to the planned two-year closure of the northbound lanes of the Pulaski Skyway that will start around March 2014, this system will synchronize traffic lights at an additional 15 intersections along US 1/9 Truck and Route 440 in Kearny Point and Jersey City.

- InSync adaptive traffic control system
- Midtown in Motion - New York City's adaptive traffic control system that employs multiple technologies. Cameras, microwave motion sensors and radio-frequency identification (RFID) E-ZPass tag readers are used as inputs as a mean to for monitoring traffic flow. The data is fed through the government-dedicated broadband wireless infrastructure to the traffic management center to be used in adaptive traffic control of the traffic lights.
- Scalable Urban Traffic Control
- Sydney Coordinated Adaptive Traffic System

RFID E-ZPass reader attached to the pole and its antenna (right) used in traffic monitoring in New York City by using vehicle re-identification method

Other Types of Control

- Failures: If power is still available, a flashing amber light is used to warn of an intersection. Methods to distinguish the main road from the secondary road (and hence right of way) include using yield (give way) signs, stop signs or a flashing red light on the secondary road as well as written signage. In some countries including Australia, the road rules outline procedures such as giving way to the right.
- Part-time operation: Some traffic lights will not operate at night or when traffic is very light. Some may only operate at particular set times (e.g. during working hours of a major factory) or only during special events such as sports or exhibitions. When not operating, the same measures as with failures are used. Part-time operation has advantages and disadvantages.
- Railroad preemption: Traffic signals are activated to coincide with the approach of a train, often where the intersection is near a rail crossing.
- Bus and Transport Priority: Traffic signals are activated to coincide with the arrival of a bus or tram along a busway, bus lane or tramway.

- Emergency Vehicles Some lights outside of fire or rescue stations have no green, as they may turn only amber and then red when fire trucks, ambulances, or other emergency vehicles or the like are exiting the station en route to an emergency.
- Speed signs are a rarely used variant to give drivers a recommended speed to approach the next traffic light in its green phase.

Active Traffic Management

Active traffic management (ATM) is the ability to dynamically manage recurrent and non-recurrent congestion based on prevailing and predicted traffic conditions. Focusing on trip reliability, it maximizes the effectiveness and efficiency of the facility. It increases throughput and safety through the use of integrated systems with new technology, including the automation of dynamic deployment to optimize performance quickly and without delay that occurs when operators must deploy operational strategies manually. ATM approaches focus on influencing travel behavior with respect to lane/facility choices and operations. ATM strategies can be deployed singularly to address a specific need such as the utilizing adaptive ramp metering to control traffic flow or can be combined to meet system-wide needs of congestion management, traveler information, and safety resulting in synergistic performance gains.

ATM strategies include:

- Variable Speed Limits.
- Temporary Shoulder Use.
- Queue Warning.
- Dynamic Merge Control.
- Adaptive Ramp Flow Control.
- Dynamic Truck Restrictions.
- Dynamic Rerouting & Traveler Information.

Used widely and successfully in Europe, ATM relies on technology to detect current conditions and automatically deploy these strategies. These strategies have proven to be very effective at delivering the lowest congestion level for the system and can be combined with managed lanes to better optimize performance.

Target Market

- Freeways or roads experiencing frequent daily congestion.
- Areas susceptible to adverse weather conditions, bottlenecks, or crashes.

It is currently in operation on the M42 motorway south-east of Birmingham and in Warwickshire. The scheme had initially been criticised by some due to possible safety and environmental concerns, however a Highways Agency report into the first six months of the scheme showed a reduction in the number of accidents from over 5 a month to 1.5 per month on average. it has now been expanded onto other roads following the initial evaluation on the M42. It is seen as a less expensive alternative to road widening.

Technology

The section of road subject to ATM is monitored by MIDAS sensor loops placed in the road every 100 metres (328 ft) (which is closer than normal) to observe traffic flows. A computerised system monitors the traffic flows and can set the best speed limit for the current flow of traffic and switch on speed limit signs mounted on gantries up to 2 kilometres (1.24 mi) before an incident. Operators can also monitor 150 CCTV cameras along the route and can control both the speed limits and information signs. Overhead variable message signs can direct drivers to use the hard shoulder during busy periods.

When the speed limit has been lowered to 50 miles per hour (80 km/h) or below the hard shoulder can be opened as an additional lane. To facilitate this and still maintain safety a series of refuge areas have been created around every 500 metres (1,640 ft) along that stretch of the road. These take the form of lay bys to the side of the hard shoulder and contain the SOS phones within them. In the event of a vehicle breaking down on the hard shoulder, operators can close it or they can close a lane to allow emergency services access to an accident. The hard shoulder is never opened on the sections under a junction between the off and on slip roads. Close to junctions use of the hard shoulder as a lane is restricted to traffic exiting or entering at that junction.

ATM involves converting the hard shoulder into a normal lane during periods of high traffic flow to expand the capacity of the road and may reduce the need to widen motorways. Similar schemes have already been implemented in Europe.

The system makes use of Automatic Number Plate Recognition cameras which to monitor traffic flows and tailor the system. Digital enforcement cameras are also mounted on the gantries and are operated by the West Midlands Police to enforce the mandatory variable speed limits.

Future Implementation

- Vehicular Ad-Hoc Network
- Intelligent Traffic-light Management

A number of highways in the United States have variable message signs and variable speed limits. The New Jersey Turnpike has been using active signage since the 1960s, though systems have evolved over time as they have been deployed in other areas of the country. A modern implementation of active traffic management was activated in 2010 using IRIS on Interstate 35W in Minneapolis, Minnesota and its southern suburbs as part of the Urban Partnership Agreement. Active lane management on I-35W is being combined with high-occupancy toll lanes and will be joined by a bus rapid transitway. An ATM scheme was deployed on 10 August 2010 in Washington.

In Washington State

The Washington State Department of Transportation (WSDOT) has completed Active Traffic Management schemes on several motorways in the Seattle metropolitan area. It is the first legally enforceable system in the United States. Failure to comply with speed limits and overhead instructions are citable offences.

ATM systems were activated on 11.6 km (7.2 mi) of the I-5 northbound carriageway in August 2010. In November 2010, ATM was expanded to 12.4 km (7.7 mi) of the SR 520 in both directions. In March 2011, ATM completed testing and began operations on 14.3 km (8.9 mi) of the I-90 in both directions.

The ATM schemes build upon WSDOT's existing arsenal of Intelligent Transport Systems (ITS) which is supported by traffic sensor loops embedded in the pavement approximately every 800 m (2640 ft) apart. On motorway sections with ATM, gantries are also spaced roughly 800 m (2640 ft) apart. The primary ATM strategies used by WSDOT is ramp metering, queue protection, hard shoulder running, junction control, and lane-specific signalling. Motorway operations in the Seattle area are conducted in the Northwest Region Traffic Management Centre in Shoreline, just north of Seattle.

Ramp Metering

WSDOT's ramp metering strategy is aimed at reducing the volume of traffic entering the motorway and has been in operation since the early 1980s. Traffic signals are operated part-time on slip roads, which are used to temporarily store traffic. The queuing traffic is then released onto the motorway one vehicle per signal cycle.

Unlike other states in the U.S., WSDOT does not have a time-of-day schedule for metering, nor does it have a rigid meter rate as it is considered too inflexible. Dedicated operators monitor traffic conditions visually through CCTV and switch the meters on and off manually. Once turned on, the meter rate is automatically determined and updated every 20 seconds using a local traffic-responsive algorithm based on fuzzy logic. The algorithm, named the Fuzzy Logic Ramp Metering algorithm, is the successor to the Bottleneck Algorithm.

The length of the queue on the slip road and the mainline occupancy immediately surrounding the slip road are fed as inputs to the algorithm, which determines a meter rate that allows as few vehicles to join the motorway mainline as possible without overflowing queuing vehicles onto nearby arterial streets. During its operation, ramp metering is fully automated. Operators will manually tune the ramp meters if necessary. They also have the ability to intervene when a malfunction occurs.

The performance of Fuzzy Logic Ramp Metering is comparable to the ALINEA algorithm used by several European agencies.

Queue Protection

Variable speed limits displayed on overhead lane control signs (LCS) above each lane are used to reduce traffic speed prior to a congestion point. Variable message signs (VMS) accompany the reduced speed limit to warn drivers of slow traffic. Speeds are determined automatically and are lowered using one or two upstream gantries depending on the size of the reduction. Intervals of 8.0 km/h (5.0 mph) and 16.1 km/h (10.0 mph) are common. Upon the end of a congestion point, speed limits are returned to the default speed. The primary purpose of queue protection is to reduce rear-end collisions.

Hard Shoulder Running and Junction Control

As part of the SR 520 Bridge Replacement and HOV scheme, WSDOT plans to implement junction control through hard shoulder running. Based on traffic conditions, ATM will open the shoulder as an auxiliary lane on the SR 520 westbound carriageway across Portage Bay Bridge, which will turn the motorway section from a 3 + 2 to a 3 + 3 dual carriageway. The additional capacity will allow traffic from the Montlake Boulevard junction more room to merge. The shoulder running can also be activated for incident management. A similar scheme is being developed for the I-5 northbound carriageway in Marysville.

Lane Specific Signalling

In the event of a lane closure due to a collision or roadwork, LCS signs will display a red X above the closed lane at the location of the incident. The gantry immediately upstream will direct motorists to merge into adjacent lanes. Drivers are allowed approximately 800 m (2640 ft), or one gantry interval, to clear the lane. Lane closures are done manually through operator intervention. Operators also have the ability to override HOV designation above HOV lanes, opening it to regular traffic if necessary.

Implications

Although WSDOT has not published data on the performance of the ATM implementation, low compliance with variable speed limits and overhead instructions is noticeable among Washington drivers. Part of the challenge is allowing drivers more time to adapt to the system, while the other part is poor enforcement from the Washington State Patrol and the absence of automated enforcement such as those used in England.

How ATM will Help?

- Improve safety through a reduction in crashes during congestion and adverse weather conditions by adjusting traffic speed and flow and providing drivers with real-time information.
- Delay onset of congestion by increasing capacity allowing traffic to flow smoothly and efficiently and improving travel time reliability.
- Provide environmental benefits through decreased emissions, noise, and fuel consumption.
- Relatively low cost to install and maintain.

Implementation Issues

Since most all ATM techniques are new to the United States, public acceptance and understanding is critical to their success. In some, additional right-of-way may be required for overhead gantries, additional lanes and shoulders, or on-ramps. Acquiring this extra space may be difficult. Many of these strategies must seamlessly work in unison with one another, often requiring automated deployment and vigilance from operators to manually adjust the system when necessary to maximize the benefit for drivers.

References

- Knoflacher, Hermann (January 2001) [2001]. Stehzeuge. Der Stau ist kein Verkehrsproblem (in German). Vienna: Böhlau. ISBN 978-3-205-98988-2
- Goddard, Haynes (July 1997). "Using Tradeable Permits to Achieve Sustainability in the World's Large Cities". Environmental and Resource Economics. Springer Netherlands. 10 (1): 63–99. doi:10.1023/A:1026444113237
- "LEDA Measure: License plate based traffic restrictions, Athens, Greece". LEDA database. Archived from the original on 2008-02-27. Retrieved 2008-04-09
- Ogilvie, David; Matt Egan; Val Hamilton; Mark Petticrew (2004-09-22). "Promoting walking and cycling as an alternative to using cars: systematic review". British Medical Journal. BMJ Publishing Group. 329 (7469): 763. doi:10.1136/bmj.38216.714560.55. PMC 520994. PMID 15385407
- Merrillees, Scott (2015). Jakarta: Portraits of a Capital 1950-1980. Jakarta: Equinox Publishing. pp. 116–7. ISBN 9786028397308
- Kerner, Boris S. (2013). "Criticism of generally accepted fundamentals and methodologies of traffic and transportation theory: A brief review". Physica A: Statistical Mechanics and its Applications. 392: 5261–5282. doi:10.1016/j.physa.2013.06.004
- "Glossary". National Agenda for Motorcycle Safety. US Department of Transportation National Highway Traffic Safety Administration/Motorcycle Safety Foundation. Retrieved 2010-09-18
- Xu, Wang (2016). "Implementation of Variable Speed Limits: Preliminary Test on Whitemud Drive, Edmonton, Canada". Journal Of Transportation Engineering. 142 (12). doi:10.1061/(ASCE)TE.1943-5436.0000895

Permissions

All chapters in this book are published with permission under the Creative Commons Attribution Share Alike License or equivalent. Every chapter published in this book has been scrutinized by our experts. Their significance has been extensively debated. The topics covered herein carry significant information for a comprehensive understanding. They may even be implemented as practical applications or may be referred to as a beginning point for further studies.

We would like to thank the editorial team for lending their expertise to make the book truly unique. They have played a crucial role in the development of this book. Without their invaluable contributions this book wouldn't have been possible. They have made vital efforts to compile up to date information on the varied aspects of this subject to make this book a valuable addition to the collection of many professionals and students.

This book was conceptualized with the vision of imparting up-to-date and integrated information in this field. To ensure the same, a matchless editorial board was set up. Every individual on the board went through rigorous rounds of assessment to prove their worth. After which they invested a large part of their time researching and compiling the most relevant data for our readers.

The editorial board has been involved in producing this book since its inception. They have spent rigorous hours researching and exploring the diverse topics which have resulted in the successful publishing of this book. They have passed on their knowledge of decades through this book. To expedite this challenging task, the publisher supported the team at every step. A small team of assistant editors was also appointed to further simplify the editing procedure and attain best results for the readers.

Apart from the editorial board, the designing team has also invested a significant amount of their time in understanding the subject and creating the most relevant covers. They scrutinized every image to scout for the most suitable representation of the subject and create an appropriate cover for the book.

The publishing team has been an ardent support to the editorial, designing and production team. Their endless efforts to recruit the best for this project, has resulted in the accomplishment of this book. They are a veteran in the field of academics and their pool of knowledge is as vast as their experience in printing. Their expertise and guidance has proved useful at every step. Their uncompromising quality standards have made this book an exceptional effort. Their encouragement from time to time has been an inspiration for everyone.

The publisher and the editorial board hope that this book will prove to be a valuable piece of knowledge for students, practitioners and scholars across the globe.

Index